Neural Networks for Identification, Prediction and Control

Duc Truong Pham and Liu Xing

Neural Networks for Identification, Prediction and Control

With 98 Figures

London Berlin Heidelberg New York
Paris Tokyo Hong Kong
Barcelona Budapest

Duc Truong Pham and Xing Liu

Intelligent Systems Laboratory
Systems Division
Cardiff School of Engineering
University of Wales
Cardiff, UK.

ISBN-13: 978-1-4471-3246-2   e-ISBN-13: 978-1-4471-3244-8
DOI: 10.007/978-1-4471-3244-8

British Library Cataloguing in Publication Data
Pham, D. T.
  Neural Networks for Identification, Prediction and Control
  I. Title II. Liu, Xing
  620.00113

Library of Congress Cataloging-in-Publication Data
Pham, D. T.
  Neural networks for identification, prediction, and control / Duc Truong Pham and Liu Xing.
    p.   cm.
  Includes bibliographical references and index.

  1. Neural networks (Computer science)   I. Liu, Xing, 1958- .   II. Title.
  QA76.87.P46 1995
  003'.363--dc20   95-12935

Typesetting: Camera ready by authors
Printed and bound at the Athenæum Press Ltd., Gateshead, Tyne and Wear

Preface

In recent years, there has been a growing interest in applying neural networks to dynamic systems identification (modelling), prediction and control. Neural networks are computing systems characterised by the ability to learn from examples rather than having to be programmed in a conventional sense. Their use enables the behaviour of complex systems to be modelled and predicted and accurate control to be achieved through training, without *a priori* information about the systems' structures or parameters.

This book describes examples of applications of neural networks in modelling, prediction and control. The topics covered include identification of general linear and non-linear processes, forecasting of river levels, stock market prices and currency exchange rates, and control of a time-delayed plant and a two-joint robot. These applications employ the major types of neural networks and learning algorithms. The neural network types considered in detail are the multilayer perceptron (MLP), the Elman and Jordan networks and the Group-Method-of-Data-Handling (GMDH) network. In addition, cerebellar-model-articulation-controller (CMAC) networks and neuromorphic fuzzy logic systems are also presented. The main learning algorithm adopted in the applications is the standard backpropagation (BP) algorithm. Widrow-Hoff learning, dynamic BP and evolutionary learning are also described.

The book is aimed at electrical, electronic, control and systems engineers concerned with system identification, prediction and control who wish to explore neural network approaches. The book does not assume a previous background in neural computing. For readers unfamiliar with the field of neural networks, Chapter 1 provides a concise review of the main existing types of neural networks and introduces frequently encountered terms. The remaining seven chapters of the book are devoted to different application areas, with essentially Chapters 2 and 3 covering modelling, Chapters 4 and 5, prediction, and Chapters 6 to 8, control.

Chapter 2 describes the use of MLPs trained with the BP algorithm for identifying dynamic systems. Both state-space and input-output modelling techniques are discussed. Simulation results obtained for various types of plants are given. A new hybrid MLP structure incorporating linear as well as non-linear neurons for modelling weakly non-linear plants is presented. As MLPs, which are feedforward neural networks, are inherently static-mapping devices, the tapped-delay-line method of feeding back to the input of an MLP previous output and input values is needed to enable them to represent dynamic processes. This method is not required for recurrent neural networks such as the Elman network detailed in Chapter 3. This chapter proposes a simple modification to the original Elman network that enables it to be trained by the standard BP algorithm to model high-order systems. The chapter shows that the proposed modification gives the standard algorithm essential characteristics of dynamic backpropagation.

Both the MLP and Elman networks have structures that are fixed by the user prior to training. Chapter 4 focuses on the GMDH network which possesses a structure that evolves during training. The chapter describes the use of the Widrow-Hoff delta rule to train GMDH networks to model and predict the behaviour of dynamic systems. Various linear and non-linear plants are identified and simulation results compared with those obtained using MLPs. A study of the application of the GMDH network to river level prediction is detailed. Other practical forecasting applications are covered in Chapter 5 which deals with financial predictions, in particular, stock market price and currency exchange rate predictions.

Chapter 6 reviews the main types of neural-network-based controllers, concentrating on those employing a neural model of the inverse dynamics of the plant. The CMAC control approach by Albus, the hierarchical approach by Kawato and the various architectures by Psaltis are described and compared with conventional adaptive control structures.

Chapter 7 treats fuzzy logic controllers (FLCs) as a special type of feedforward neural network. An FLC shares the same layered feedforward configuration as an MLP and has input-encoding facilities, similar to those in a CMAC controller, which gives it good noise rejection abilities. The chapter describes the use of the genetic algorithm (GA), which is a stochastic optimisation algorithm inspired by natural evolution processes, to train FLCs to control a plant with input time delay and a non-linear plant.

Apart from the stock market price example given in Chapter 5, all systems considered in the book so far are single-input-single-output (SISO) systems. Chapter 8 illustrates the neural control of MIMO systems and presents an inverse-model-based system for controlling a two-joint SCARA robot. The main controller is a partially-recurrent neural network of the type proposed by Jordan, but incorporating feedback from the hidden-layer neurons to the state-layer neurons as in an Elman network.

In addition to the main chapters, the book contains six appendices. Appendices A to C respectively cover background material on conventional approaches to identification, prediction and control, on fuzzy logic theory and on genetic algorithms. This summary material is provided for convenient referencing by the reader. Appendices D to F list computer programs written in C for system identification using the MLP (Appendix D) and modified Elman network (Appendix E) and for prediction using the GMDH network (Appendix F). These computer programs were employed to obtain many of the simulation results presented in the book. The listings are supplied to help the reader explore the use of these types of neural networks in identification and prediction tasks.

D.T. Pham
X. Liu

Acknowledgements

Most of the material presented in this book has originated from research in the authors' Intelligent Systems Laboratory at Cardiff. The authors gratefully acknowledge the help of laboratory members, in particular, Dr D. Karaboga and Dr S.J. Oh whose work features in Chapters 7 and 8. Other team members have generously given their time checking drafts of sections of the book for technical and typographical errors. They include Mr R. Alcock, Mr A.B. Chan, Dr P.H. Channon, Mr J.K.C. Cheung, Dr S.S. Dimov, Dr P.R. Drake, Mr N.R. Jennings, Mr G.G. Jin, Mrs X. Ji, Mr B.J. Peat, Mr D.P. Richards, Mr H. Rowlands, Dr S. Sagiroglu, Mr M.F. Sukkar, Dr A. Wani and Mr S. Yildirim. Mrs L.M. McCarthy edited and printed the final manuscript and is thanked for her assistance.

The authors would also like to thank Mr N. Pinfield, Mrs I. Mowbray, Mr N.G. Wilson and Mr R. Dobbing, their colleagues at Springer-Verlag London, for their help with the production of the book.

The book was written with the partial financial support of the Higher Education Funding Council for Wales to whom the authors are indebted.

Finally, they wish to acknowledge the permission of Computational Mechanics Publications, Taylor & Francis, the IEEE and Cambridge University Press to use material previously published by them as follows:-

- Chapter 1: material from Pham, D.T. (1994) Neural networks in engineering, *Proc. 9th Int Conf on Artificial Intelligence in Engineering*, Rzevski, G., Adey, R.A. and Russell, D.W. (eds), Malvern, PA, July 1994, 3-36.
- Chapter 4: material from Pham, D.T. and Liu, X. (1994) Modelling and prediction using GMDH networks of Adalines with nonlinear preprocessors, *Int. J. Systems Science*, 25(11), 1743-1759.

- Chapter 7: material from Pham, D.T. and Karaboga, D. (1993) Design of neuromorphic fuzzy controllers, *Proc. IEEE-SMC Conf., Systems Engineering in the Service of Human,* Le Touquet, France, Oct. 1993, **4**, 103-108.
- Chapter 8: material from Pham, D.T. and Oh, S. J. (1994) Adaptive control of a robot using neural networks, *Robotica*, **12**, 553-561.

Contents

Chapter 1 Artificial Neural Networks

Artificial neural networks are computational models of the brain. There are many types of neural networks representing the brain's structure and operation with varying degrees of sophistication. This chapter provides an introduction to the main types of networks and presents examples of each type.

1.1 Types of Neural Networks

Neural networks generally consist of a number of interconnected processing elements (PEs) or neurons. How the inter-neuron connections are arranged and the nature of the connections determine the structure of a network. How the strengths of the connections are adjusted or trained to achieve a desired overall behaviour of the network is governed by its learning algorithm. Neural networks can be classified according to their structures and learning algorithms.

1.1.1 Structural Categorisation

In terms of their structures, neural networks can be divided into two types: feedforward networks and recurrent networks.

Feedforward networks: In a feedforward network, the neurons are generally grouped into layers. Signals flow from the input layer through to the output layer via unidirectional connections, the neurons being connected from one layer to the next, but not within the same layer. Examples of feedforward networks include the multi-layer perceptron (MLP) [Rumelhart and McClelland, 1986], the learning vector quantization (LVQ) network [Kohonen, 1989], the cerebellar model articulation control (CMAC) network [Albus, 1975a] and the group-method of data handling (GMDH) network [Hecht-Nielsen, 1990]. Feedforward networks can most naturally perform static mappings between an input space and an output space: the output at a given instant is a function only of the input at that instant.

Recurrent networks: In a recurrent network, the outputs of some neurons are fedback to the same neurons or to neurons in preceding layers. Thus, signals can flow in both forward and backward directions. Examples of recurrent networks include the Hopfield network [Hopfield, 1982], the Elman network [Elman, 1990] and the Jordan network [Jordan, 1986]. Recurrent networks have a dynamic memory: their outputs at a given instant reflect the current input as well as previous inputs and outputs.

1.1.2 Learning Algorithm Categorisation

Neural networks are trained by two main types of learning algorithms: supervised and unsupervised learning algorithms. In addition, there exists a third type, reinforcement learning, which can be regarded as a special form of supervised learning.

Supervised learning: A supervised learning algorithm adjusts the strengths or weights of the inter-neuron connections according to the difference between the desired and actual network outputs corresponding to a given input.   Thus, supervised learning requires a *teacher* or *supervisor* to provide desired or target output signals.   Examples of supervised learning algorithms include the delta rule [Widrow and Hoff, 1960], the generalised delta rule or backpropagation algorithm [Rumelhart and McClelland, 1986] and the LVQ algorithm [Kohonen, 1989].

Unsupervised learning: Unsupervised learning algorithms do not require the desired outputs to be known. During training, only input patterns are presented to the neural network which automatically adapts the weights of its connections to cluster the input patterns into groups with similar features.   Examples of unsupervised learning algorithms include the

Kohonen [Kohonen, 1989] and Carpenter-Grossberg Adaptive Resonance Theory (ART) [Carpenter and Grossberg, 1988] competitive learning algorithms.

Reinforcement learning: As mentioned before, reinforcement learning is a special case of supervised learning. Instead of using a teacher to give target outputs, a reinforcement learning algorithm employs a critic only to evaluate the goodness of the neural network output corresponding to a given input. An example of a reinforcement learning algorithm is the genetic algorithm (GA) [Holland, 1975; Goldberg, 1989].

1.2 Example Neural Networks

This section briefly describes the example neural networks and associated learning algorithms cited previously.

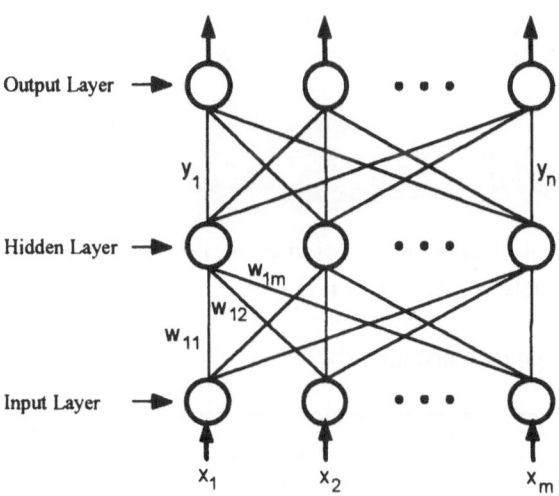

Figure 1.1 (a) A multi-layer perceptron

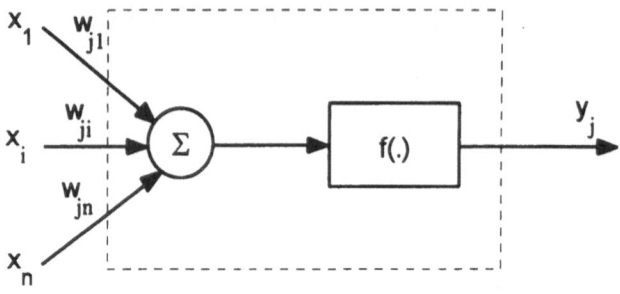

Figure 1.1 (b) Details of a neuron.

1.2.1 Multi-layer Perceptron (MLP)

MLPs are perhaps the best known type of feedforward networks. Figure 1.1(a) shows an MLP with three layers: an input layer, an output layer and an intermediate or hidden layer. Neurons in the input layer only act as buffers for distributing the input signals x_i to neurons in the hidden layer. Each neuron j (Figure 1.1(b)) in the hidden layer sums up its input signals x_i after weighting them with the strengths of the respective connections w_{ji} from the input layer and computes its output y_j as a function f of the sum, viz.

$$y_j = f\left(\sum w_{ji} x_i\right) \tag{1.1}$$

f can be a simple threshold function or a sigmoidal, hyperbolic tangent or radial basis function (see Table 1.1).

Table 1.1 Activation functions

Type of Functions	Functions
Linear	$f(s) = s$
Threshold	$f(s) = \begin{cases} +, & if \ \ s > s_t \\ -, & otherwise \end{cases}$
Sigmoid	$f(s) = 1/(1 + exp(-s))$
Hyperbolic tangent	$f(s) = (1 - exp(-2s))/(1 + exp(2s))$
Radial basis function	$f(s) = exp(-s^2/\beta^2)$

The output of neurons in the output layer is computed similarly.

The backpropagation (BP) algorithm, a gradient descent algorithm, is the most commonly adopted MLP training algorithm. It gives the change Δw_{ji} n the weight of a connection between neurons i and j as follows:-

$$\Delta w_{ji} = \eta \, \delta_j \, x_i \tag{1.2}$$

where η is a parameter called the learning rate and δ_j is a factor depending on whether neuron j is an output neuron or a hidden neuron. For output neurons,

$$\delta_j = \left(\frac{\partial f}{\partial net_j} \right)\left(y_j^{(t)} - y_j \right) \tag{1.3}$$

and for hidden neurons,

$$\delta_j = \left(\frac{\partial f}{\partial net_j} \right)\sum_q w_{qj}\,\delta_q \tag{1.4}$$

In Equation (1.3), net_j is the total weighted sum of input signals to neuron j and $y_j^{(t)}$ is the target output for neuron j.

As there are no target outputs for hidden neurons, in Equation (1.4), the difference between the target and actual output of a hidden neuron j is replaced by the weighted sum of the δ_q terms already obtained for neurons q connected to the output of j. Thus, iteratively, beginning with the output layer, the δ term is computed for neurons in all layers and weight updates determined for all connections. The weight updating process can take place after the presentation of each training pattern (pattern-based training) or after the presentation of the whole set of training patterns (batch training). In either case, a training epoch is said to have been completed when all training patterns have been presented once to the MLP.

For all but the most trivial problems, several epochs are required for the MLP to be properly trained. A commonly adopted method to speed up the training is to add a "momentum" term to Equation (1.2) which effectively lets the previous weight change influence the new weight change, viz:

$$\Delta w_{ji}(k+1) = \eta \delta_j x_i + \mu \Delta w_{ji}(k) \tag{1.5}$$

where $\Delta w_{ji}(k+1)$ and $\Delta w_{ji}(k)$ are weight changes in epochs $(k+1)$ and (k) respectively and μ is the "momentum" coefficient.

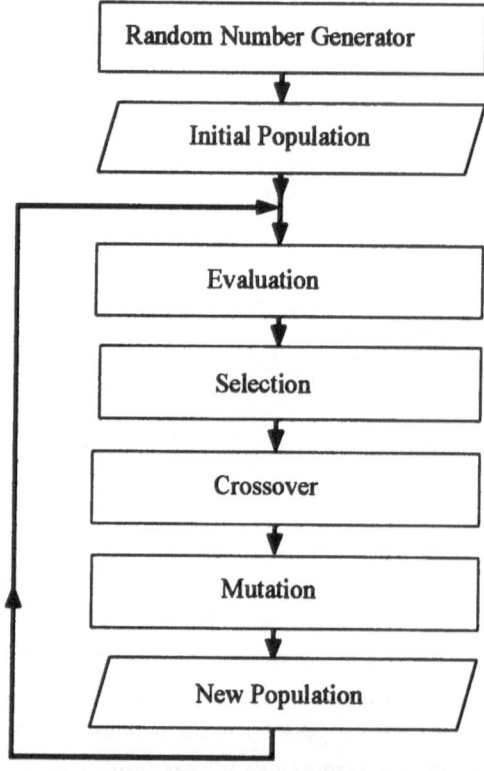

Figure 1.2 (a) A simple genetic algorithm

Another learning algorithm suitable for training MLPs is the GA (see Figure 1.2(a)). This is an optimisation algorithm based on evolution principles. The weights of the connections are considered genes in a chromosome. The goodness or fitness of the chromosome is directly related to how well trained the MLP is. The algorithm starts with a randomly generated population of chromosomes and applies genetic operators to create new and fitter populations. The most common genetic operators are the selection, crossover and mutation operators. The selection operator chooses chromosomes from the current population for reproduction. Usually, a biased selection procedure is adopted which favours the fitter chromosomes. The crossover operator (Figure 1.2(b)) creates two new chromosomes from two existing chromosomes by cutting them at a random position and exchanging the parts following the cut. The mutation operator (Figure 1.2(c)) produces a new chromosome by randomly changing the genes of an existing chromosome. Together, these operators simulate a guided random search method which can eventually yield the optimum set of weights to minimise the differences between the actual and target outputs of the neural network. Further details of genetic algorithms can be found in Appendix C.

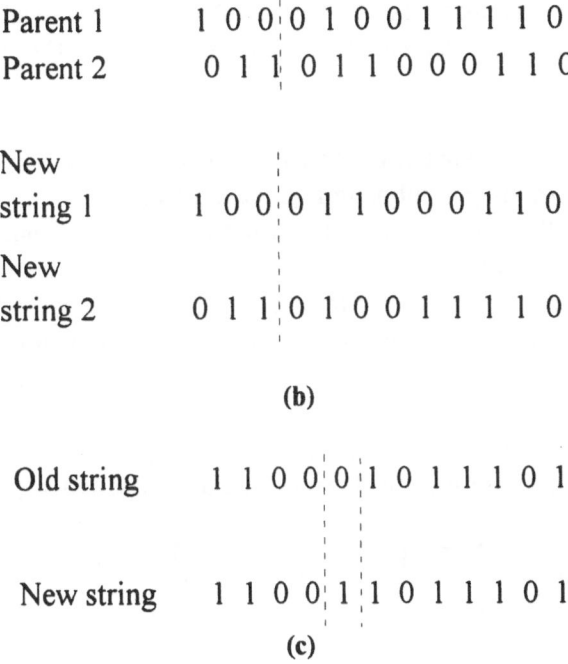

Parent 1   1 0 0 0 1 0 0 1 1 1 1 0
Parent 2   0 1 1 0 1 1 0 0 0 1 1 0

New
string 1   1 0 0 0 1 1 0 0 0 1 1 0

New
string 2   0 1 1 0 1 0 0 1 1 1 1 0

(b)

Old string   1 1 0 0 0 1 0 1 1 1 0 1

New string   1 1 0 0 1 1 0 1 1 1 0 1

(c)

Figure 1.2 (b) Crossover operation; **(c)** Mutation operation.

1.2.2 Learning Vector Quantization (LVQ) network

Figure 1.3 shows an LVQ network which comprises three layers of neurons: an input buffer layer, a hidden layer and an output layer. The network is fully connected between the input and hidden layers and partially connected between the hidden and output layers, with each output neuron linked to a different cluster of hidden neurons. The weights of the connections between the hidden and output neurons are fixed to 1. The weights of the input-hidden neuron connections form the components of *reference* vectors (one reference vector is assigned to each hidden neuron). They are modified during the training of the network. Both the hidden neurons (also known as Kohonen neurons) and the output neurons have binary outputs. When an input pattern is supplied to the network, the hidden neuron whose reference vector is closest to the input pattern is said to win the competition for being activated and thus allowed to produce a "1". All other hidden neurons are forced to produce a "0". The output neuron connected to the cluster of hidden neurons that contains the winning neuron also emits a "1" and all other output neurons a "0". The output neuron that produces a "1" gives the class of the input pattern, each output neuron being dedicated to a different class. The simplest LVQ training procedure is as follows:-

(i)   initialise the weights of the reference vectors;
(ii)   present a training input pattern to the network;
(iii) calculate the (Euclidean) distance between the input pattern and each reference vector;
(iv)   update the weights of the reference vector that is closest to the input pattern, that is, the reference vector of the winning hidden neuron. If the latter belongs to the cluster connected to the output neuron in the class that the input pattern is known to belong to, the reference vector is brought closer to the input pattern. Otherwise, the reference vector is moved away from the input pattern;

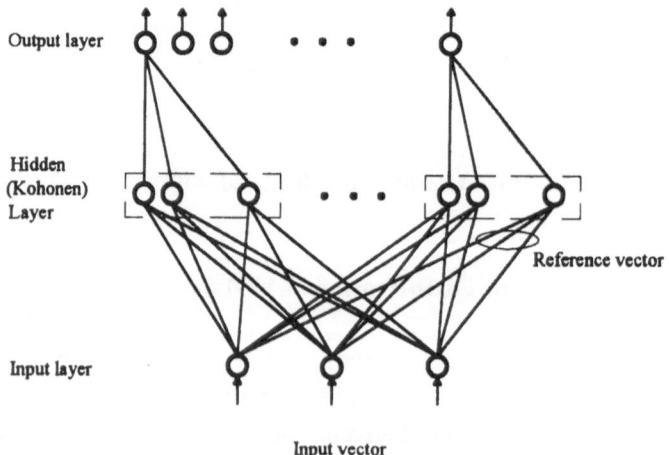

Figure 1.3 Learning Vector Quantization network

(v)   return to (ii) with a new training input pattern and repeat the procedure until all training patterns are correctly classified (or a stopping criterion is met).

For other LVQ training procedures, see for example [Pham and Oztemel, 1994].

1.2.3 CMAC network

CMAC (Cerebellar Model Articulation Control) [Albus, 1975a, 1975b, 1979a, 1979b; An et al 1994] can be considered a supervised feedforward neural network with the characteristics of a fuzzy associative memory. A basic CMAC module is shown in Figure 1.4.

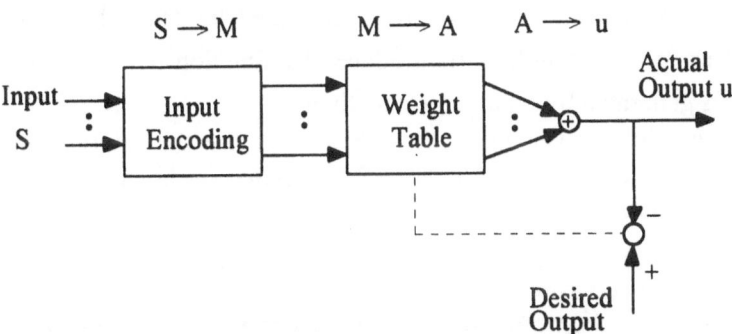

Figure 1.4 A basic CMAC module

CMAC consists of a series of mappings:

$$S \xrightarrow{\quad e \quad} M \xrightarrow{\quad f \quad} A \xrightarrow{\quad g \quad} u \qquad (1.6)$$

where

$$S = \{\text{input vectors}\}$$
$$M = \{\text{intermediate variables}\}$$
$$A = \{\text{association cell vectors}\}$$

u = output of CMAC ≡ $h(\mathbf{S})$

$h \equiv \mathbf{g} \circ \mathbf{f} \circ \mathbf{e}$

(a). Input encoding (S → M mapping)

The S → M mapping is a set of submappings, one for each input variable:

$$\mathbf{S} \rightarrow \mathbf{M} = \begin{bmatrix} s_1 \rightarrow \mathbf{m}_1 \\ s_2 \rightarrow \mathbf{m}_2 \\ \vdots \\ s_n \rightarrow \mathbf{m}_n \end{bmatrix} \tag{1.7}$$

The range of s_i is coarsely discretised using the quantising functions $q_1, q_2, \dots, q_k$. Each function divides the range into k intervals. The intervals produced by function q_{j+1} are offset by one kth of the range compared to their counterparts produced by function q_j. $\mathbf{m}_i$ is a set of k intervals generated by q_1 to q_k respectively.

An example is given in Figure 1.5 to illustrate the internal mappings within a CMAC module. The S → M mapping is shown in the leftmost part of the figure.

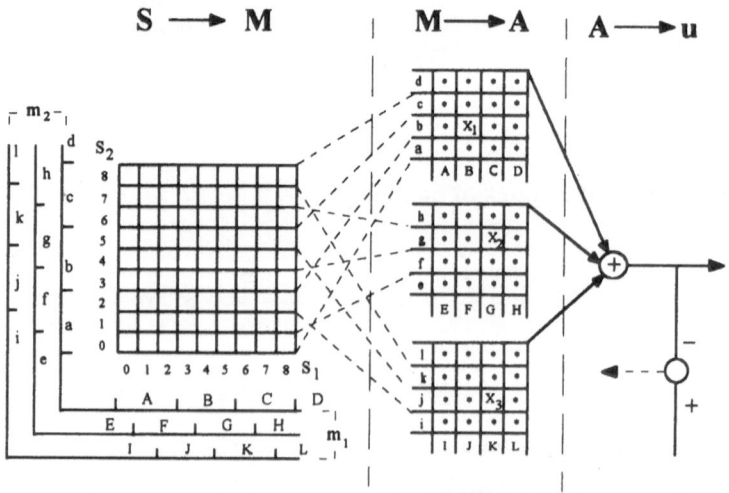

Figure 1.5 Internal mappings within a CMAC module

In Figure 1.5, two input variables s_1 and s_2 are represented with unity resolution in the range of 0 to 8. The range of each input variable is described using three quantising functions. For example, the range of s_1 is described by functions q_1, q_2, and q_3. q_1 divides the range into intervals A, B, C and D. q_2 gives intervals E, F, G, and H and q_3 provides intervals I, J, K and L. That is,

$$q_1 = \{A, B, C, D\}$$
$$q_2 = \{E, F, G, H\}$$
$$q_3 = \{I, J, K, L\}$$

For every value of s_1, there exists a set of elements, $\mathbf{m}_1$, which are the intersection of the functions q_1 to q_3, such that the value of s_1 uniquely defines set $\mathbf{m}_1$ and vice versa. For example, value $s_1 = 5$ maps to set $m_1 = \{B, G, K\}$ and vice versa. Similarly, value $s_2 = 4$ maps to set $\mathbf{m}_2 = \{b, g, j\}$ and vice versa.

The $S \rightarrow M$ mapping gives CMAC two advantages: the first is that a single precise variable s_i can be transmitted over several imprecise information channels. Each channel carries only a small part of the information of s_i. This increases the reliability of the information transmission. The other advantage is that small changes in the value of s_1 have no influence on most of the elements in $\mathbf{m}_i$. This leads to the property of input generalisation which is important in an environment where random noise exists.

(b). Address computing ($M \rightarrow A$ mapping)

A is a set of address vectors associated with weight tables. A is obtained by combining the elements of $\mathbf{m}_i$. For example, in Figure 1.5, the sets $\mathbf{m}_1 = \{B, G, K\}$ and $\mathbf{m}_2 = \{b, g, j\}$ are combined to give the set of elements $\mathbf{A} = \{a_1, a_2, a_3\} = \{Bb, Gg, Kj\}$.

(c). Output computing ($A \rightarrow u$ mapping)

This mapping involves looking up the weight tables and adding the contents of the addressed locations to yield the output of the network. The following formula is employed:

$$u = \sum_i w_i(a_i) \qquad (1.8)$$

That is, only the weights associated with the addresses a_i in A are summed. For this given example, these weights are:

$$w(Bb) = x_1$$
$$w(Gg) = x_2$$
$$w(Kj) = x_3$$

Thus the output is:

$$u = x_1 + x_2 + x_3 \qquad (1.9)$$

Training a CMAC module consists of adjusting the stored weights. Assuming that f is the function that CMAC has to learn, the following training steps could be adopted:

(i) Select a point S in the input space and obtain the current output u corresponding to S ;

(ii) Let $\bar{u}$ be the desired output of CMAC, that is, $\bar{u} = f(S)$;

(iii) If $|\bar{u} - u| \leq \xi$, where ξ is an acceptable error, then do nothing; the desired value is already stored in CMAC. However, if $|\bar{u} - u| > \xi$, then add to every weight which contributed to u the quantity

$$\Delta = \alpha \frac{\bar{u} - u}{|A|} \qquad (1.10)$$

where $|A|$ = the number of weights which contributed to u and α is the learning rate.

1.2.4 Group Method of Data Handling (GMDH) Network

Figure 1.6 shows a GMDH network and the details of one of its neurons. Unlike the feedforward neural networks previously described which have a fixed structure, a GMDH network has a structure which grows during training. Each neuron in a GMDH network usually has two inputs x_1 and x_2 and produces an output y that is a quadratic combination of these inputs, viz.

$$y = w_o + w_1 x_1 + w_2 x_1^2 + w_3 x_1 x_2 + w_4 x_2^2 + w_5 x_2 \qquad (1.11)$$

Training a GMDH network consists of configuring the network starting with the input layer, adjusting the weights of each neuron, and increasing the number of layers until the accuracy of the mapping achieved with the network deteriorates.

The number of neurons in the first layer depends on the number of external inputs available. For each pair of external inputs, one neuron is used.

Training proceeds with presenting an input pattern to the input layer and adapting the weights of each neuron according to a suitable learning algorithm, such as the delta rule (see for example [Pham and Liu, 1994]), viz.

$$W_{k+1} = W_k + \alpha \frac{X_k}{|X_k|^2} \left(y_k^d - W_k^T X_k \right) \qquad (1.12)$$

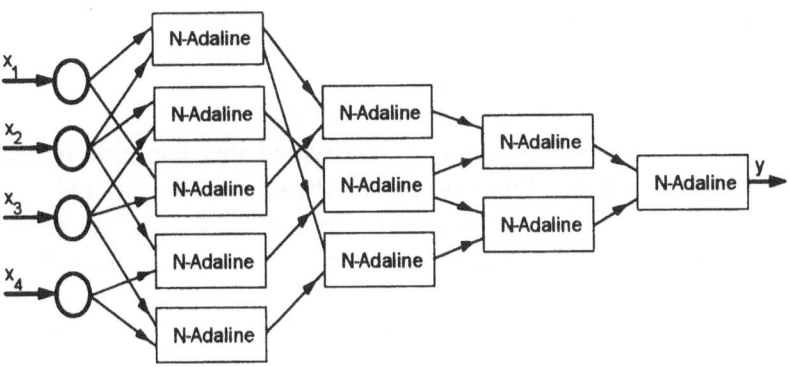

Note: Each GMDH neuron is an N-Adaline, which is an Adaptive Linear Element with a nonlinear preprocessor

Figure 1.6 (a) A trained GMDH network

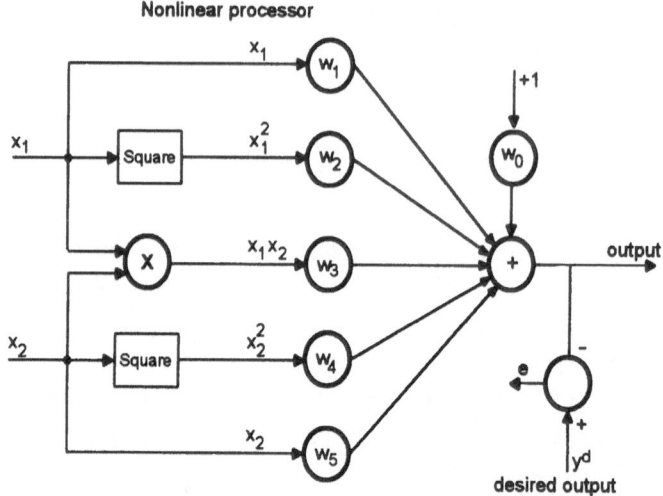

Figure 1.6 (b) Details of a GMDH Neuron

where W_k, the weight vector of a neuron at time k, and X_k the modified input vector to the neuron at time k, are defined as

$$W_k = \begin{bmatrix} w_0 & w_1 & w_2 & w_3 & w_4 & w_5 \end{bmatrix}^T \qquad (1.13)$$

$$X_k = \begin{bmatrix} 1 & x_1 & x_1^2 & x_1 x_2 & x_2^2 & x_2 \end{bmatrix}^T \qquad (1.14)$$

and y_k^d is the desired network output at time k.

Note that, for this description, it is assumed that the GMDH network only has one output. Equation (1.12) shows that the desired network output is presented to each neuron in the input layer and an attempt is made to train each neuron to produce that output. When the sum of the mean square errors S_E over all the desired outputs in the training data set for a given neuron reaches the minimum for that neuron, the weights of the neuron are frozen and its training halted. When the training has ended for all neurons in a layer, the training for the layer stops. Neurons that produce S_E values below a given threshold when another set of data (known as the selection data set) is presented to the network are selected to grow the next layer. At each stage, the smallest S_E value achieved for the selection data set is recorded. If the smallest S_E value for the current layer is less than that for the previous layer (that is, the accuracy of the network is improving), a new layer is generated, the size of which depends on the number of neurons just selected. The training and selection processes are repeated until the S_E value deteriorates. The best neuron in the immediately preceding layer is then taken as the output neuron for the network.

1.2.5 Hopfield network

Figure 1.7 shows one version of a Hopfield network. This network normally accepts binary and bipolar inputs (+1 or -1). It has a single "**layer**" of neurons, each connected to all the others, giving it a recurrent structure, as mentioned earlier. The *training* of a Hopfield network takes only one step, the weights w_{ij} of the network being assigned directly as follows:-

$$w_{ij} = \begin{cases} \dfrac{1}{N}\displaystyle\sum_{c=1}^{P} x_i^c\, x_j^c\,, & i \neq j \\[4mm] 0 & ,\quad i = j \end{cases} \qquad (1.15)$$

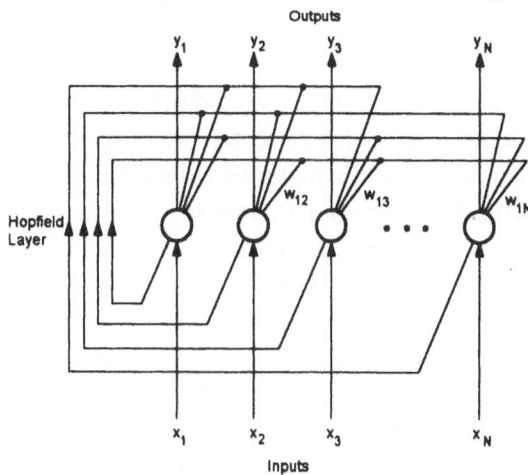

Figure 1.7 A Hopfield network

where w_{ij} is the connection weight from neuron i to neuron j, and x_i^c (which is either +1 or -1) is the ith component of the training input pattern for class c, P the number of classes and N the number of neurons (or the number of components in the input pattern). Note from Equation (1.15) that $w_{ij}=w_{ji}$ and $w_{ii}=0$, a set of conditions that guarantee the stability of the network. When an unknown pattern is input to the network, its outputs are initially set equal to the components of the unknown pattern, viz.

$$y_i(0) = x_i \qquad 1 \le i \le N \tag{1.16}$$

Starting with these initial values, the network iterates according to the following equation until it reaches a minimum *energy* state, i.e. its outputs stabilise to constant values:-

$$y_i(k+1) = f\left[\sum_{j=1}^{N} w_{ij}\, y_i(k)\right] \qquad 1 < i \le N \tag{1.17}$$

where f is a hard limiting function defined as

$$f(x) = \begin{cases} -1 & x < 0 \\ 1 & x > 0 \end{cases} \tag{1.18}$$

1.2.6 Elman and Jordan nets

Figures 1.8 (a) and (b) show an Elman net and a Jordan net, respectively. These networks have a multi-layered structure similar to the structure of MLPs. In both nets, in addition to an ordinary hidden layer, there is another special hidden layer sometimes called the context or state layer. This layer receives feedback signals from the ordinary hidden layer (in the case of an Elman net) or from the output layer (in the case of a Jordan net). The Jordan net also has connections from each neuron in the context layer back to itself. With both nets, the outputs of neurons in the context layer, are fed forward to the hidden layer. If only the forward connections are to be adapted and the feedback connections are preset to constant values, these networks can be considered ordinary feedforward networks and the BP algorithm used to train them. Otherwise, a GA could be employed

[Pham and Karaboga, 1993b; Karaboga, 1994]. For improved versions of the Elman and Jordan nets, see [Pham and Liu, 1992; Pham and Oh, 1992].

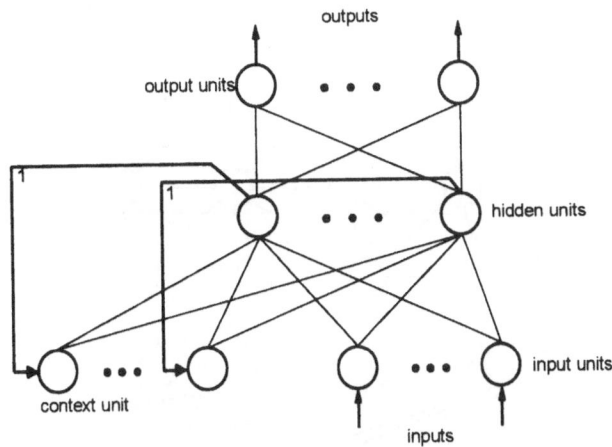

Figure 1.8 (a) An Elman network

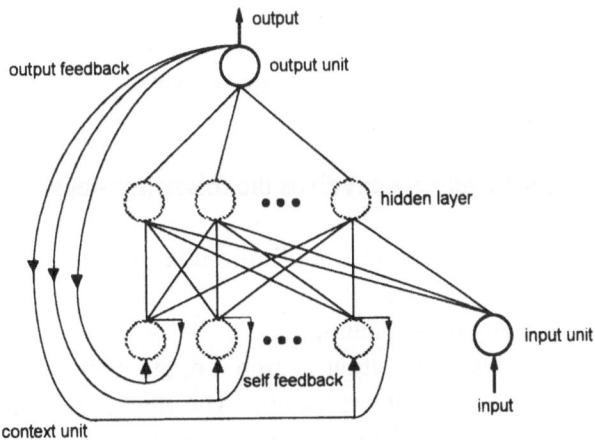

Figure 1.8 (b) A Jordan network

1.2.7 Kohonen network

A Kohonen network or a self-organising feature map has two layers, an input buffer layer to receive the input pattern and an output layer (see

Figure 1.9). Neurons in the output layer are usually arranged into a regular two-dimensional array. Each output neuron is connected to all input neurons. The weights of the connections form the components of the reference vector associated with the given output neuron.

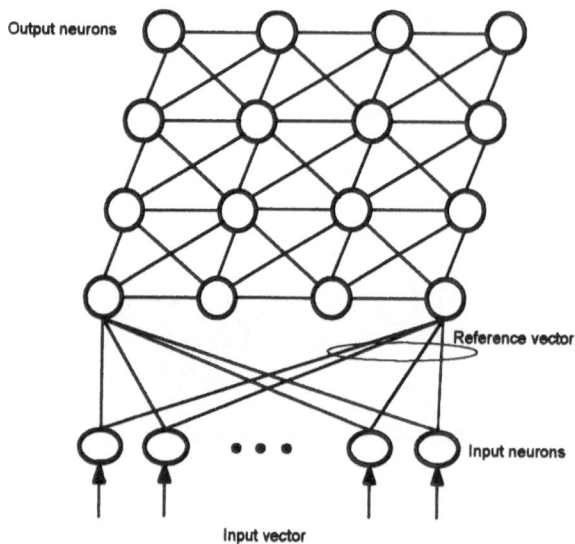

Figure 1.9 A Kohonen network

Training a Kohonen network involves the following steps:

(i)   initialise the reference vectors of all output neurons to small random values;

(ii) present a training input pattern;

(iii) determine the winning output neuron, i.e. the neuron whose reference vector is closest to the input pattern. The Euclidean distance between a reference vector and the input vector is usually adopted as the distance measure;

(iv) update the reference vector of the winning neuron and those of its neighbours. These reference vectors are brought closer to the input vector. The adjustment is greatest for the reference vector of the winning neuron and decreased for reference vectors of neurons further away. The size of the neighbourhood of a neuron is reduced as training proceeds until, towards the end of training, only the reference vector of a winning neuron is adjusted.

In a well-trained Kohonen network, output neurons that are close to one another have similar reference vectors. After training, a labelling procedure is adopted where input patterns of known classes are fed to the network and class labels are assigned to output neurons that are activated by those input patterns. As with the LVQ network, an output neuron is activated by an input pattern if it wins the competition against other output neurons, that is, if its reference vector is closest to the input pattern.

1.2.8 ART network

There are different versions of the ART network. Figure 1.10 shows the ART-1 version for dealing with binary inputs. Later versions, such as ART-2, can also handle continuous-valued inputs.

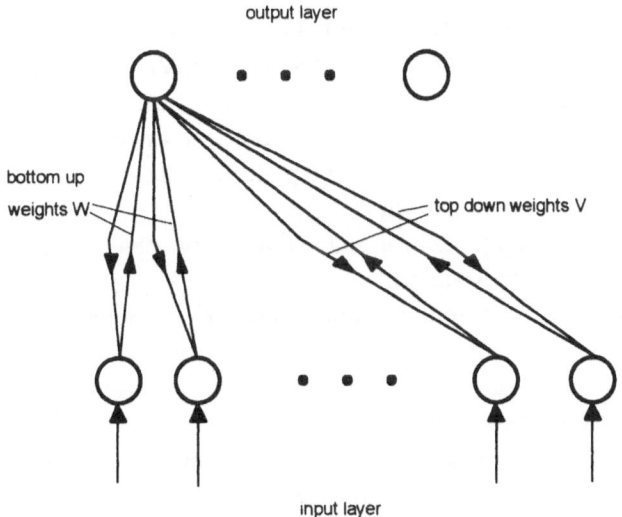

Figure 1.10 An ART-1 network

As illustrated in Figure 1.10, an ART-1 network has two layers, an input layer and an output layer. The two layers are fully interconnected, the connections are in both the forward (or bottom-up) direction and the feedback (or top-down) direction. The vector W_i of weights of the bottom-up connections to an output neuron i forms an exemplar of the class it represents. All the W_i vectors constitute the long-term memory of the network. They are employed to select the winning neuron, the latter again being the neuron whose W_i vector is most similar to the current input pattern. The vector V_i of the weights of the top-down connections from an output neuron i is used for *vigilance* testing, that is, determining whether an input pattern is sufficiently close to a stored exemplar. The vigilance vectors V_i form the short-term memory of the network. V_i and W_i are related in that W_i is a normalised copy of V_i, viz.

$$W_i = \frac{V_i}{\varepsilon + \sum V_{ji}} \tag{1.19}$$

where ε is a small constant and V_{ji}, the jth component of V_i (i.e. the weight of the connection from output neuron i to input neuron j).

Training an ART-1 network occurs continuously when the network is in use and involves the following steps:-

(i)   initialise the exemplar and vigilance vectors W_i and V_i for all output neurons, setting all the components of each V_i to 1 and computing W_i according to Equation (1.19). An output neuron with all its vigilance weights set to 1 is known as an *uncommitted* neuron in the sense that it is not assigned to represent any pattern classes;

(ii) present a new input pattern x;

(iii) enable all output neurons so that they can participate in the competition for activation;

(iv) find the winning output neuron among the competing neurons, i.e. the neuron for which $x.W_i$ is largest; a winning neuron can be an uncommitted neuron as is the case at the beginning of training or if there are no better output neurons;

(v)   test whether the input pattern x is sufficiently similar to the vigilance vector V_i of the winning neuron. Similarity is measured by the fraction r of bits in x that are also in W_i, viz.

$$r = \frac{x.V_i}{\sum x_i} \tag{1.20}$$

x is deemed to be sufficiently similar to V_i if r is at least equal to *vigilance threshold* ρ ($0 < \rho \leq 1$);

(vi) go to step (**vii**) if $r \geq \rho$ (i.e. there is *resonance*); else disable the winning neuron temporarily from further competition and go to step (**iv**) repeating this procedure until there are no further enabled neurons;

(vii) adjust the vigilance vector V_i of the most recent winning neuron by logically ANDing it with x, thus deleting bits in V_i that are not also in x; compute the bottom-up exemplar vector W_i using the new V_i according to Equation (1.19); activate the winning output neuron;

(viii) go to step (**ii**).

The above training procedure ensures that if the same sequence of training patterns is repeatedly presented to the network, its long-term and short-term memories are unchanged (i.e. the network is *stable*). Also, provided there are sufficient output neurons to represent all the different classes, new patterns can always be learnt, as a new pattern can be assigned to an uncommitted output neuron if it does not match previously stored exemplars well (i.e. the network is *plastic*).

1.3 Summary

This chapter has presented the main types of existing neural networks and has described examples of each type. For an overview of the different systems engineering applications of these neural networks, see (Pham, 1994) for example.

References

Albus, J. S. (1975a) A new approach to manipulator control: cerebellar model articulation control (CMAC), *Trans. ASME, J. of Dynamics Syst., Meas. and Contr.*, **97**, 220-227.

Albus, J. S. (1975b) Data storage in the cerebellar model articulation controller (CMAC), *Trans. ASME, J. of Dynamics Syst., Meas. and Contr.*, **97**, 228-233.

Albus, J. S. (1979a) A model of the brain for robot control, *Byte*, 54-95.

Albus, J. S. (1979b) Mechanisms of planning and problem solving in the brain, *Math. Biosci.*, **45**, 247-293.

An, P.E., Brown, M., Harris, C.J., Lawrence, A.J., Moore, C.J. (1994) Associative memory neural networks: adaptive modelling theory, software implementations and graphical user, *Engng. Appli. Artif. Intell.*, 7(1), 1-21.

Carpenter, G.A. and Grossberg, S. (1988) The ART of adaptive pattern recognition by a self-organising neural network, *Computer*, March 1988, 77-88.

Elman, J.L. (1990) Finding structure in time, *Cognitive Science*, **14**, 179-211.

Goldberg, D. (1989) *Genetic Algorithms in Search, Optimisation and Machine Learning*, Reading, MA: Addison-Wesley.

Hecht-Nielsen, R. (1990) *Neurocomputing*, Reading, MA: Addison-Wesley.

Holland, J.H. (1975) *Adaptation in Natural and Artificial Systems*, Ann Arbor, MI: University of Michigan Press.

Hopfield, J.J. (1982) Neural networks and physical systems with emergent collective computational abilities, *Proceedings of the National Academy of Sciences*, **79**, 2554-2558.

Jordan, M.I. (1986) Attractor dynamics and parallelism in a connectionist sequential machines, *Proceedings of the 8th Annual Conference of the Cognitive Science Society*, 531-546.

Karaboga, D. (1994) *Design of Fuzzy Logic Controllers Using Genetic Algorithms*, PhD thesis, University of Wales, Cardiff, UK.

Kohonen, T. (1989) Self-Organising and Associative Memory (3rd ed.), Berlin: Springer-Verlag.

Pham, D.T. (1994) Neural networks in engineering, *Proc. 9th Int. Conf. on Artificial Intelligence in Engineering*, Malvern, PA, July 1994, 3-36.

Pham, D.T. and Karaboga, D. (1993) Dynamic system identification using recurrent neural networks and genetic algorithms, *Proc. 9th Int. Conf. on Mathematical and Computer Modelling*, San Francisco, July 1993, in press.

Pham, D.T. and Liu, X. (1994) Modelling and prediction using GMDH networks of Adalines with nonlinear preprocessors, *Int. J. Systems Science*, **25**(11), 1743-1759.

Pham, D.T and Oh, S.J. (1992) A recurrent backpropagation neural network for dynamic system identification, *Journal of Systems Engineering*, **2**(4), 213-223.

Pham, D.T. and Liu, X. (1992) Dynamic system modelling using partially recurrent neural networks, *Journal of Systems Engineering*, **2**(2), 90-97.

Pham, D.T. and Oztemel, E. (1994) Control chart pattern recognition using learning vector quatization networks, *Int. J. Production Research*, 32(3), 721-729.

Rumelhart, D. and McClelland, J. (1986) *Parallel distributed processing: exploitations in the micro-structure of cognition*, volumes 1 and 2, Cambridge: MIT Press.

Widrow, B. and Hoff, M.E. (1960) Adaptive switching circuits, *Proc. 1960 IRE WESCON Convention Record*, Part 4, IRE, New York, 96-104

Chapter 2 Dynamic System Identification Using Feedforward Neural Networks

A dynamic system can be described by two types of models: input-output models and state-space models. This chapter describes the use of feedforward neural networks to learn to act as both types of models.

2.1 Dynamic System Descriptions

2.1.1 Input-Output Model

An input-output model describes a dynamic system based on input and output data. In the discrete-time domain, an input-output model can be of the NARMAX type [Chen and Billings, 1990] or the parametric Hammerstein type [Iserman et al, 1992]. An input-output model assumes that the new system output can be predicted by the past inputs and outputs of the system. If a system is further supposed to be deterministic, time invariant, single-input-single-output (SISO), the input-output model becomes:

$$y_p(k) = f(y_p(k-1), y_p(k-2), ..., y_p(k-n) \\ u(k-1), u(k-2), ..., u(k-m)) \tag{2.1}$$

where $[u(k), y_p(k)]$ represents the input-output pair of the system at time k. Positive integers n and m are respectively the number of past outputs (also called the order of the system) and the number of past inputs. In practice m is usually smaller than or equal to n. f can be a static nonlinear function which maps the past inputs and outputs to a new output.

If a system is linear, f is a linear function and Equation (2.1) can be rewritten as:

$$y_p(k) = a_1 y_p(k-1) + a_2 y_p(k-2), \ldots, + a_n y_p(k-n)$$
$$+ b_1 u(k-1) + b_2 u(k-2), \ldots, + b_m u(k-m)$$

$$(2.2)$$

where $a_i (i = 1, 2, \ldots, n)$ and $b_i (i = 1, 2, \ldots, m)$ are real constants.

2.1.2 State-Space Model

A dynamic system can also be described by a state-space model. The state-space model of an nth-order multi-input-multi-output (MIMO) time-invariant nonlinear dynamic system is as follows:

$$\mathbf{x}(k+1) = \Phi[\mathbf{x}(k), \mathbf{u}(k)]$$
$$\mathbf{y}(k) = \Psi[\mathbf{x}(k)]$$

$$(2.3)$$

where $\mathbf{x}(k) = [x_1(k), x_2(k), \ldots, x_n(k)]^T$ is the n-component state vector of the system, $\mathbf{u}(k) = [u_1(k), u_2(k), \ldots, u_r(k)]^T$ is the system input vector, and $\mathbf{y}(k) = [y_1(k), y_2(k), \ldots, y_m(k)]^T$ is the system output vector. Φ and Ψ are static nonlinear mappings. If the system is linear, Equation (2.3) becomes:

$$\mathbf{x}(k+1) = \mathbf{A}\mathbf{x}(k) + \mathbf{B}\mathbf{u}(k)$$
$$\mathbf{y}(k) = \mathbf{C}\mathbf{x}(k)$$

$$(2.4)$$

where $\mathbf{A}, \mathbf{B}$, and $\mathbf{C}$ are $(n \times n)$, $(n \times r)$, and $(m \times n)$ matrices.

2.2 Identification Based on System Inputs and Outputs

Feedforward neural networks do not have dynamic memory. They have been shown to be able to approximate arbitrary static functions [Funahashi, 1989; Hornik, 1991]. The task of system identification is essentially to find suitable mappings which can approximate the mappings implied in a dynamic system.

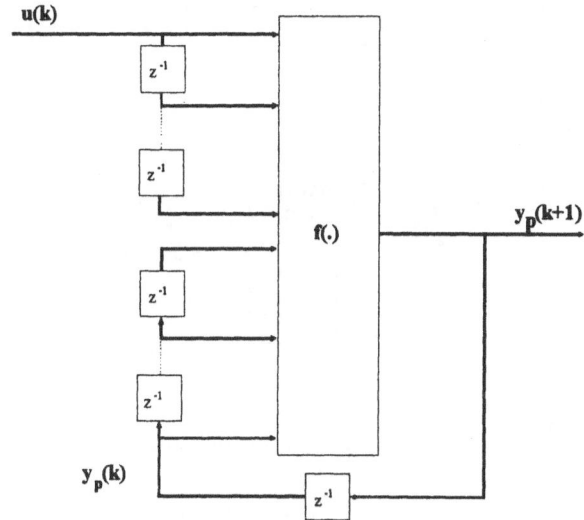

Figure 2.1 Input-output model

Equation (2.1) can be represented by the block diagram shown in Figure 2.1. It can be seen from Figure 2.1 that when input-output data are used, the dynamic system is defined by function f and integers m and n. If m and n are given, the only task is to find f. f does not change with time for time-invariant systems.

Since a feedforward neural network can represent static mappings, it can be employed to approximate f. According to the configuration of an identification system, there can be two identification structures: parallel structure (Figure 2.2) and series-parallel structure (Figure 2.3).

In the parallel structure, the network and the system receive the same external inputs; the outputs of the system are not used as inputs to the network. The system and the network are two independent processes which share the same external inputs. Their outputs do not interfere with one another.

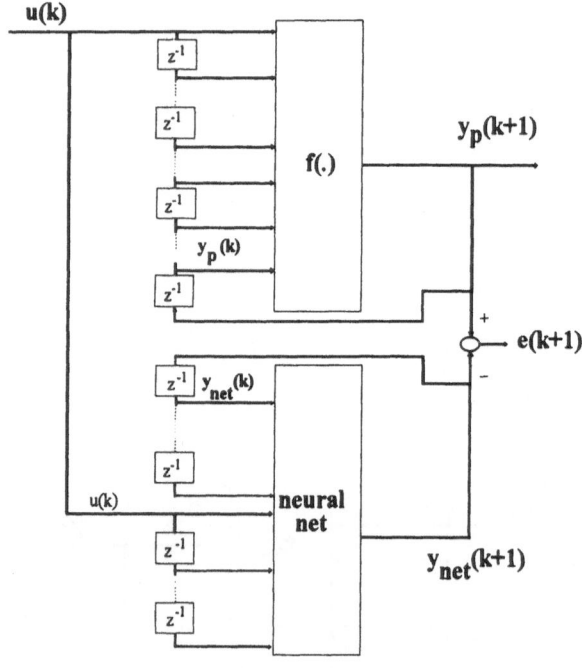

Figure 2.2 Parallel identification structure

In the series-parallel structure, again the network and the system receive the same external inputs, but the output of the system is part of the inputs to the network.   The system and the network are no longer two independent processes. The dynamic behaviour of the network is affected by the system.

When the structures of Figures 2.2 and 2.3 are employed for identification, the system is assumed to be bounded-input bounded-output (BIBO) stable in the presence of an input. However, if the parallel structure is employed, it cannot be guaranteed that the learning process of the weights will converge or the error between the output of the system and that of the network will tend to zero.

The neural networks in both the parallel and series-parallel identification structures require $(n+m)$ input units and one output unit.

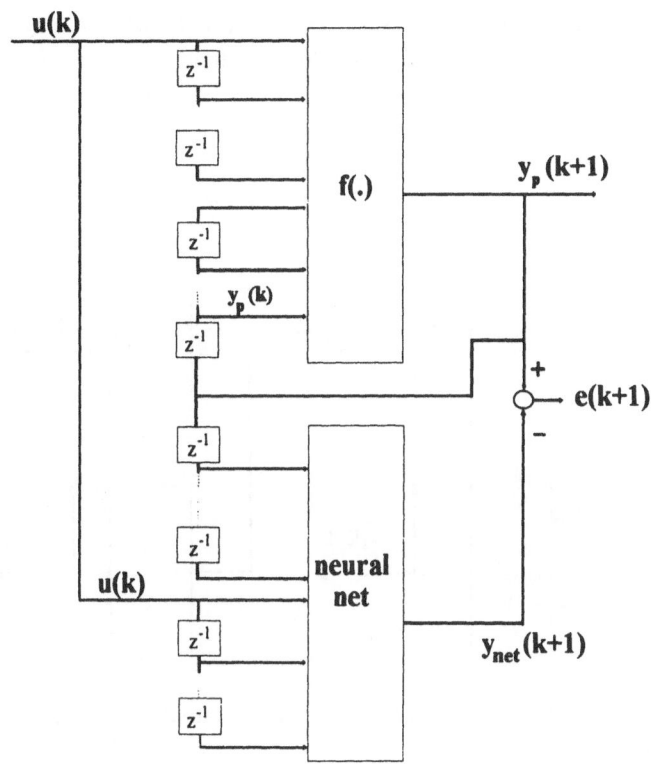

Figure 2.3 Series-parallel identification structure

2.3 Identification Based on Measurable System States

In Equation (2.3), $\Phi(.)$ and $\Psi(.)$ are arbitrary functions which may or may not be linear. They are the mappings for a neural network to approximate. One is mapping $\Phi(.)$ which maps the state $x(k)$ and input $u(k)$ into the new state $x(k+1)$. The other is mapping $\Psi(.)$ which transforms the state $x(k)$ into the output $y(k)$. When neural networks are used to identify the systems described by Equation (2.3), two assumptions are made: (i) all the system states are measurable; (ii) the system is stable.

Based on the above assumptions, the architecture depicted in Figure 2.4 can be used for system identification. Figure 2.4 shows that two networks are

needed. Neural network 1 has the system inputs and states as its input
signals. Thus the input buffer layer of neural network 1 has $(n + r)$ units.
Moreover, since neural network 1 has the predicted new system states as its
output signals, its output layer needs n units. The numbers of hidden layers
and hidden units may be decided according to accuracy requirements.
Similarly, neural network 2 needs n input units and m output units.

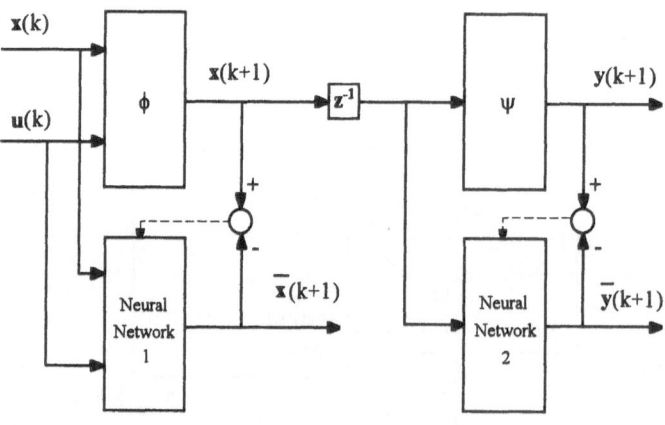

Figure 2.4 Identification based on measurable system states

2.4 Input-Output Model Identification

Recent research has demonstrated that neural networks are effective for the
identification of dynamic systems. The work reported by Narendra and
Parthasarathy [Narendra and Parthasarathy, 1990] concentrated on the
identification of nonlinear discrete dynamic systems that are difficult for
conventional theories to deal with. Nonlinear systems represented by input-
output models are classified into groups according to how the nonlinearities
occur in the model and the structure of the identification system is of the
series-parallel type. A number of highly nonlinear systems have been
successfully identified by neural networks with sigmoidal nonlinear
processing elements (PEs). The studies described by Yamada and Yabuta
[Yamada and Yabuta, 1990] are mainly concerned with the identification of
transfer functions and inverse transfer functions of the input-output type.
Both linear and nonlinear systems have been considered. When a linear

system is to be identified, only linear PEs are used in the neural network identifier. Nonlinear PEs are adopted for the hidden layer of the neural network when the system to be identified is nonlinear.

The above-mentioned studies suggest that neural networks with linear PEs are sufficient for linear system identification, while nonlinear PEs are needed only if the systems to be identified are nonlinear. Because the identification problem is regarded as that of finding the mapping between two sets of variables (the past system inputs and outputs and the new system outputs), the task of the neural network identifier is effectively one of function approximation. It has been shown that a neural network with sigmoidal PEs in its hidden layer(s) can universally approximate any continuous functions [Cybenko, 1989; Funahashi, 1989; Hornik, 1991]. However, some of the open research questions in this area are: for linear systems, are linear networks or nonlinear networks more appropriate? Is it useful to have a network which has both linear and nonlinear PEs? What systems are more appropriate for such a network to identify? These questions will be addressed in the following which presents the results of computer simulations of the identification of linear and nonlinear systems using different types of neural networks.

2.4.1 Simulation Investigations

As mentioned above, the literature shows that when feedforward networks are employed for identification, they have a layered structure with either only linear or only nonlinear PEs in their hidden layer(s). However, according to common intuition, linear networks should be better at identifying linear systems than nonlinear networks and this leads to the idea of hybrid networks. This section aims at examining the performance of different networks in identifying linear and nonlinear systems.

System models employed: Three models are employed throughout the simulations. They represent systems with different types of nonlinearities. The first model is a second-order linear plant with transfer function:

$$G(s) = \frac{1-e^{-Ts}}{s} \frac{\omega}{(s+a)^2 + \omega^2} \tag{2.5}$$

where the first part of $G(s)$, $\dfrac{1-e^{-Ts}}{s}$, represents the sample-and-hold device which is adopted here in consideration of the practical situation in digital systems, the sampling period $T = 0.1$ second, and the system characteristic parameters $a = 1.0$, and $\omega = 2\pi/2.5$. After discretisation, system (2.5) can be described by a second-order difference equation, viz:

$$y(k) = A_1 y(k-1) + A_2 y(k-2) + B_1 u(k-1) + B_2 u(k-2) \qquad (2.6)$$

where $A_1 = 1.752821$, $A_2 = -0.818731$, $B_1 = 0.011698$, $B_2 = 0.010942$.

Equation (2.6) can be extended to represent a nonlinear system with different types of nonlinearities by purposely introducing nonlinear terms into it, as shown in the following:

$$\begin{aligned} y(k) &= A_1 y(k-1) + A_2 y(k-2) + A_3 y^2(k-2) \\ &\quad + B_1 u(k-1) + B_2 u(k-2) + B_3 u^2(k-1) \end{aligned} \qquad (2.7)$$

It can easily be seen that by varying parameters A_3 and B_3 (within stability limits) systems with different degrees of nonlinearity can be obtained. Equation (2.7) degenerates to Equation (2.6) if A_3 and B_3 are all zero.

The second system model adopted for simulations is that of a simple pendulum swinging through small angles [Bogoliubov and Mitropolsky, 1961]. The discrete-time description of the system is:

$$\begin{aligned} y(k) &= (2 - \frac{\lambda T}{ML^2}) y(k-1) + (\frac{\lambda T}{ML^2} - 1 - \frac{gT^2}{L}) y(k-2) \\ &\quad + \frac{gT^2}{6L} y^3(k-2) - \frac{T^2}{ML^2} u(k-2) \end{aligned} \qquad (2.8)$$

where M is the mass of the pendulum, L the length, g the acceleration due to gravity, λ is the friction coefficient, y the angle of deviation from the vertical position, and u the external force exerted on the pendulum. The parameters used in this model are as follows:

$T = 0.2$ s, $\qquad\qquad$ $g = 9.8$ m/s^2

$\lambda = 1.2$ kgm^2/s, $\qquad\qquad$ $M = 1.0$ kg,

$L = 0.5$ m.

Replacing the parameters with their values in equation (2.8) gives:

$$y(k) = A_1 y(k-1) + A_2 y(k-2) + A_3 y^3(k-2) + B_1 u(k-2) \qquad (2.9)$$

where $A_1 = 1.04$, $A_2 = -0.824$, $A_3 = 0.130667$, $B_1 = -0.16$. Because only the term associated with A_3 is nonlinear, in the simulations, A_3 is again varied over a range of values from 0.01 to 0.11 to enable the system to possess different degrees of nonlinearity.

The last system is a system with "strong" nonlinearity adopted from [Narendra and Parthasarathy, 1990]:

$$y(k) = \frac{y(k-1)}{1 + y^2(k-1)} + u^3(k-1) \qquad (2.10)$$

No variable parameter is introduced to this model in the simulations.

Neural network identifiers: Three types of neural networks are used for identification. All the neural networks are multi-layer perceptrons with an input layer (buffer layer), a single hidden layer (with biases), and an unbiased linear output layer. The neural networks differ from one another in their hidden layers.

Type 1 - linear network (Figure 2.5(a)): all the PEs in the hidden layer have linear activation functions (the output of a PE is the sum of its inputs);

Type 2 - nonlinear network (Figure 2.5(b)): the PEs in the hidden layer have hyperbolic tangent activation functions defined as follows:

$$f(x) = \frac{e^x - e^{-x}}{e^x + e^{-x}} \qquad (2.11)$$

where x is the total input to a PE and $f(x)$ is the output of that PE.

Type 3 - hybrid network (Figure 2.5(c)): half of the PEs in the hidden layer are linear and the other half have hyperbolic tangent activation functions.

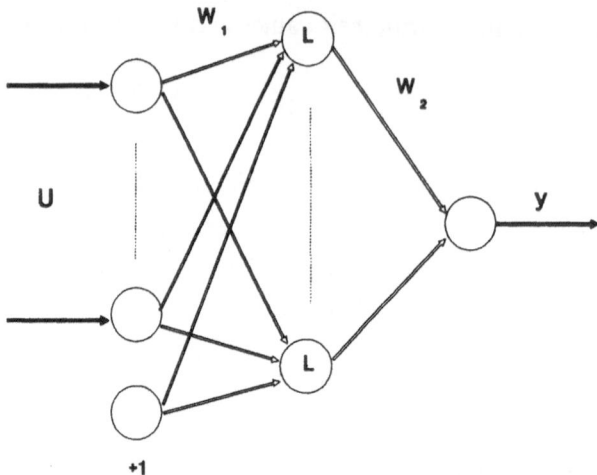

Figure 2.5 (a) Linear network (L: linear activation functions)

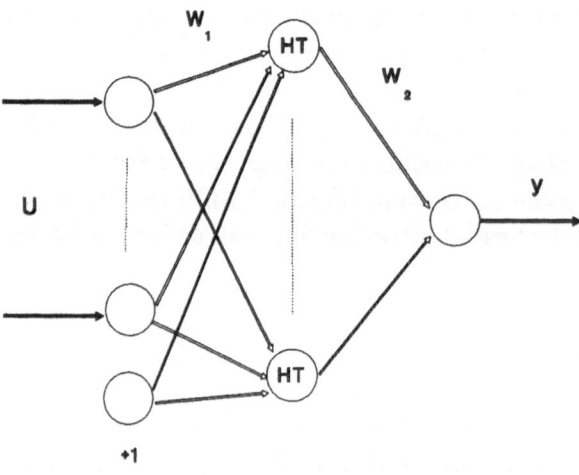

Figure 2.5 (b) Nonlinear network (HT: hyperbolic tangent functions)

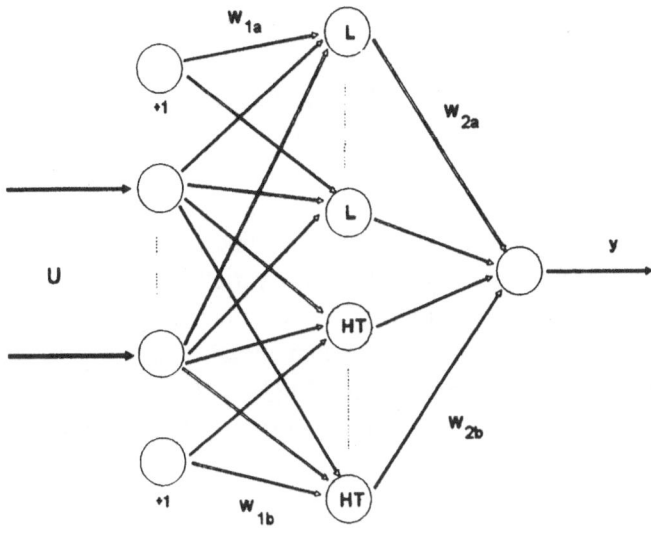

Figure 2.5 (c) Hybrid network (L: linear activation functions; HT: hyperbolic tangent functions)

Simulation set-up: The neural network is trained to emulate the dynamic behaviour of the system. The network has a fixed topological structure, i.e., the number of layers and the number of PEs in each layer are not modifiable during training. According to the models used, the networks have $(n + m)$ input PEs, a pre-selected number of hidden PEs, and an output PE. The training database consists of 400 data and during training this database is presented to the network a given number of times (i.e. training by epoch is adopted). The root-mean-square error of one training epoch, defined as follows

$$RMS \text{ error} = \sqrt{\frac{\sum_{k=1}^{400}[y(k) - y_{net}(k)]^2}{400}} \tag{2.12}$$

is used as a criterion to evaluate the performance of the given network identifiers, where $y(k)$ is the output of the model output, $y_{net}(k)$ is the output of the network. This criterion is an indication of the speed of convergence of the learning process.

Other parameters such as learning rate and momentum are also fixed before training started.

2.4.2 Results

Linear system identification: The model used in this group of simulations is given by Equation (2.6). The linear network, nonlinear network, and hybrid network are all employed in the simulations. All the networks have 4 PEs in the input layer, 6 PEs in the hidden layer (the hidden PEs can be linear, nonlinear, or half linear and half nonlinear) and one PE in the output layer. The learning rate is adopted as a variable parameter ranging from 0.1 to 0.5 in increments of 0.1.   The other parameters adopted are: momentum = 0.1; seed for the random initial weights = 1; seed for the random input $u(k)$ = 1; amplitude of $u(k)$ = $(a^2 + \omega^2)/\omega$ = 2.911162; number of training epochs = 50. The results are shown in Figure 2.6.

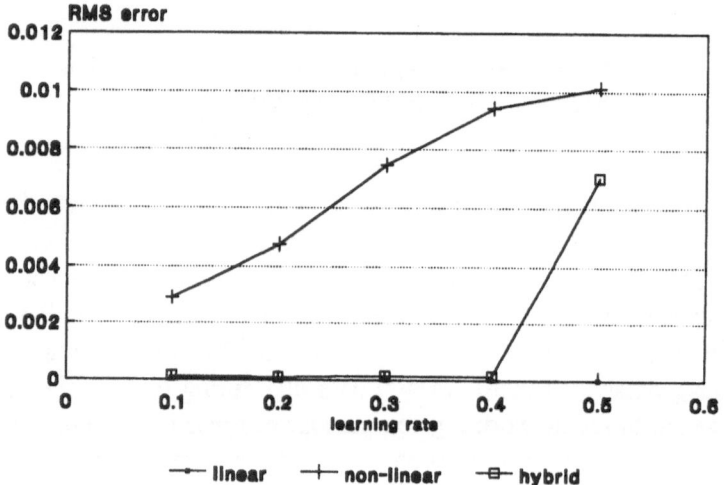

Figure 2.6 Results for linear system (Equation 2.6)

Nonlinear system identification: Linear networks are incapable of identifying nonlinear systems. Therefore, only the nonlinear network and the hybrid network are used to identify the nonlinear systems.   The parameters related to the network structures and training are the same as those used in the linear system identification.

The system models used in the simulations are given by Equations (2.7), (2.9), and (2.10). For Equation (2.7), two groups of simulations are conducted. In the first group of simulations the parameter A_3 is changed over the range 0.0 to 0.09 (with $B_3 = 0$). In the second group of simulations, the variable parameter is B_3 (with $A_3 = 0$). The range of B_3 is 0.0 to 0.1. The other parameters are the same as those used in the case of linear system identification. The simulation results are indicated in Figures 2.7(a)-(b).

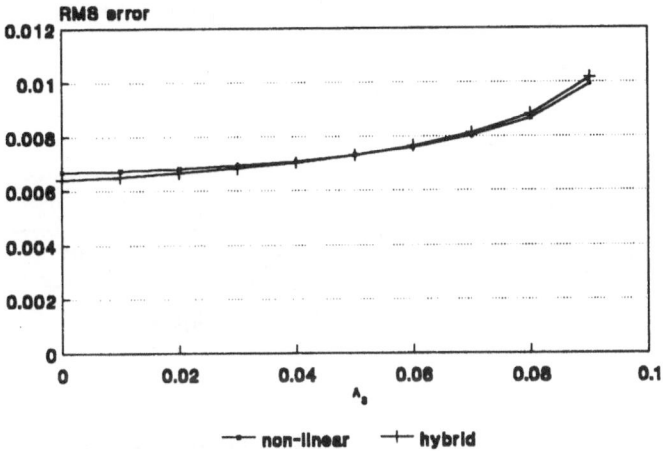

Figure 2.7 (a) Results for nonlinear system (Equation 2.7): A_3 variable.

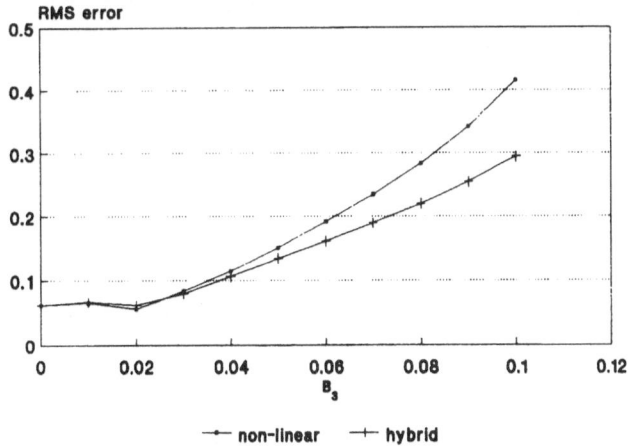

Figure 2.7 (b) Results for nonlinear system (Equation 2.7): B_3 variable.

Another group of simulations is carried out for the system represented by Equation (2.9) where A_3 is a variable parameter in the range 0.01 to 0.11 (the amplitude of $u(k)$ in this case is 2). The results are shown in Figure 2.8.

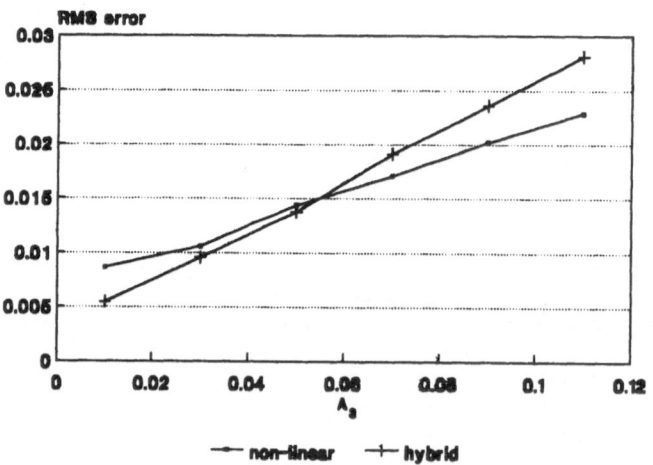

Figure 2.8 Results for nonlinear system (Equation (2.9))

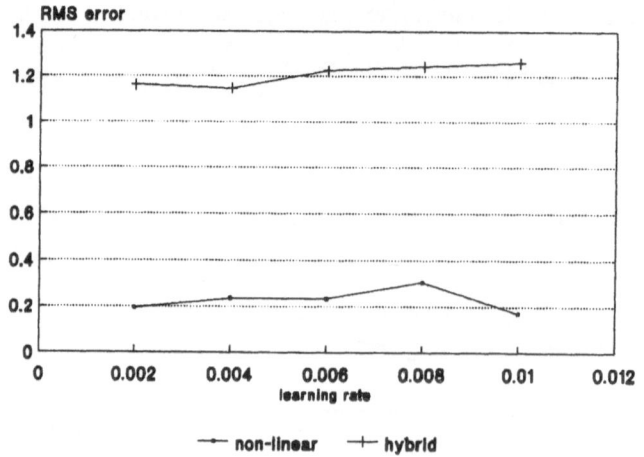

Figure 2.9 Results for nonlinear system (Equation (2.10))

Finally, the results for the "strongly" nonlinear system (Equation (2.10)) with the learning rate as variable parameter are depicted in Figure 2.9 (the amplitude for the random training input $u(k)$ in this case is also 2).

2.5 State-Space Model Identification

Simulations are carried out to test the effectiveness of the identification structure based on the state-space model. Although, as can be seen from Equation (2.3), the identification method should also work for higher-order and nonlinear systems, for simplicity, the simulations are limited to a second-order linear plant.

2.5.1 Plant Model and Identification Architecture

Considering a second-order plant as shown in Figure 2.10 [Ogata, 1970]:

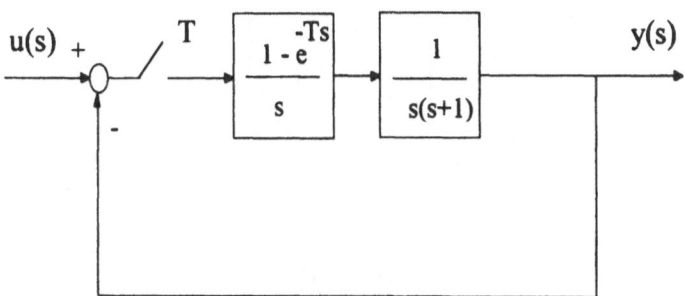

Figure 2.10 A second-order linear system

The open-loop continuous transfer function of the plant is

$$G(s) = \frac{1 - e^{-Ts}}{s^2(\tau + 1)} \tag{2.13}$$

If $T = 0.2$s and $\tau = 1$s, then the corresponding difference equation is

$$y(k+1) = 1.8y(k) - 0.837y(k-1) + 0.019u(k) + 0.018u(k-1)$$

(2.14)

The state-space equivalent of the above equation is

$$\begin{bmatrix} x_1(k+1) \\ x_2(k+1) \end{bmatrix} = \begin{bmatrix} 0 & 1 \\ -0.837 & 1.8 \end{bmatrix} \begin{bmatrix} x_1(k) \\ x_2(k) \end{bmatrix} + \begin{bmatrix} 0.019 \\ 0.052 \end{bmatrix} u(k)$$

$$y(k) = x_1(k)$$

(2.15)

In this case, the identifier is as shown in Figure 2.11. The neural net architecture is depicted in Figure 2.12.

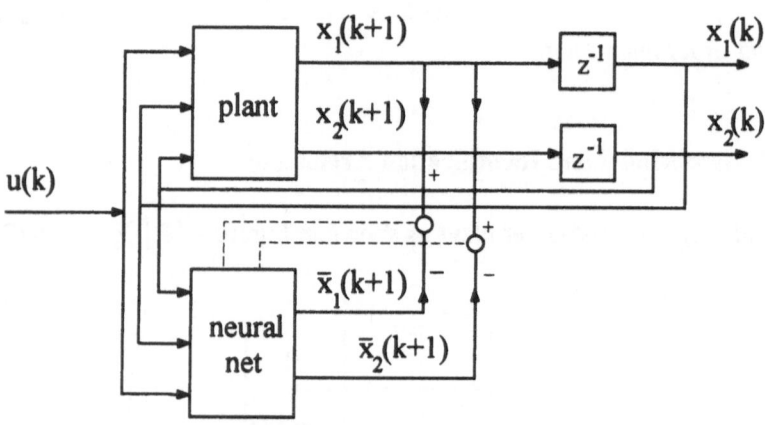

Figure 2.11 Identification scheme

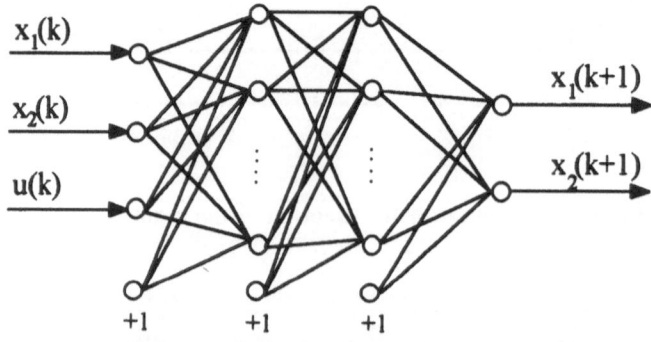

Figure 2.12 Neural network details

2.5.2 Simulations

The neural net is trained using the architecture of Figure 2.11 and the backpropagation algorithm with momentum term. In this case, the parameters are: input buffer: 3 units; first hidden layer: 12 units; second hidden layer: 12 units; output layer: 2 units; learning coefficient: 0.15; momentum: 0.2; training epochs: 450; input u(k): random signal with amplitude uniformly distributed in the interval [-1, 1]; activation functions: linear. Figure 2.13(a) and (b) respectively illustrate the responses of the identifier and plant to step and sinusoidal input signals.

2.6 Discussion

2.6.1 Hybrid Networks

From Figure 2.6 it can be seen that the linear network is the fastest in learning to identify the linear system. After only 50 epochs of training, the RMS errors approached almost zero. The performance of the hybrid network is very close to that of the linear network

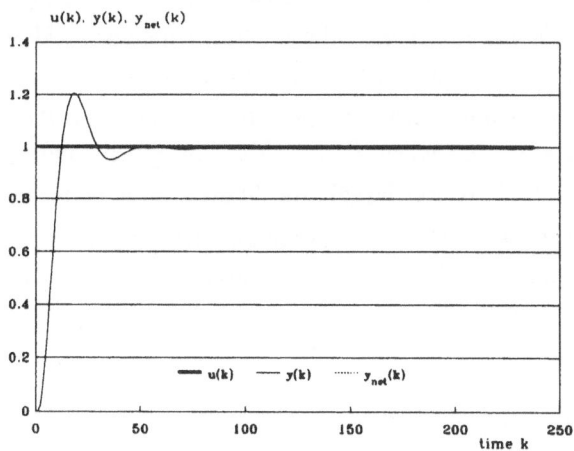

Figure 2.13 (a) Step response of state space identifier

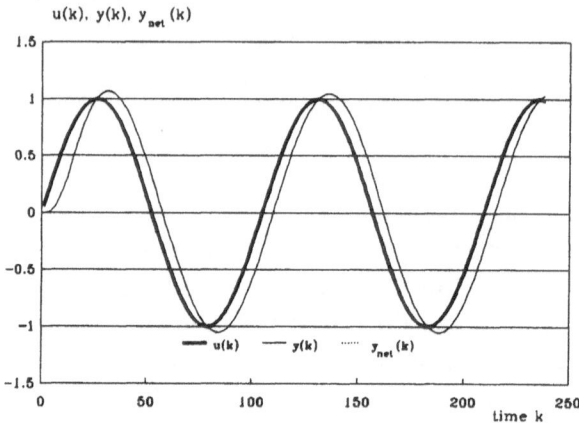

Figure 2.13 (b) Sinusoidal response of state space identifier

when the learning rate is between 0.1 and 0.4. Although both the nonlinear network and the hybrid network can achieve small training RMS errors, the nonlinear network always produce larger errors. This indicates that the linear network is the best network for linear system identification and the hybrid network is better than the nonlinear network at that task.

Figures 2.7(a) and 2.8 illustrate that when the nonlinearity in a nonlinear system is weak ($A_3 < 0.05$ in Figure 2.7(a), $A_3 \leq 0.05$ in Figure 2.8), the hybrid network learns faster than the nonlinear network. However, when the nonlinearity becomes stronger ($A_3 > 0.05$ in Figure 2.7(a), $A_3 > 0.06$ in Figure 2.8), the nonlinear network performs better. On the other hand, Figure 2.7(b) shows that when B_3 is within the given range (the training cannot proceed if B_3 is too large), the hybrid network is always faster. Finally, for a system as shown in Equation (2.10) which has very strong nonlinearity, Figure 2.9 shows that the nonlinear network is always better than the hybrid network.

2.6.2 Input-output versus State-space Modelling

The results for the state space based identification (Figures 13(a) and (b)) show that the neural nets can successfully identify the input-output behaviour of the system. Although two hidden layers have been employed in the state-space modelling example, it is sufficient to use only one hidden layer since the network has linear activation functions and the overall matrix representing the network input-output relationship can be obtained by multiplying the matrices of the two stages corresponding to the two hidden layers. If the system states are available and some of the states are outputs, then only one network is needed. This network requires fewer input units than the network used in the input-output structure. For example, six input units are necessary for a third-order system in the input-output structure and for the state-space structure, only four input units are required. Also, in the input-output structure, more historical data are needed to predict the new output, but the state-space structure needs only the present data. Finally, the state-space model can represent a wider range of dynamic systems than the input-output model, as pointed out in [Srinivasan et al, 1994].

2.7 Analysis of the Hybrid Network

As illustrated in [Harber and Unbehauen, 1990; Iserman et al, 1992], the input-output mapping function $f(.)$ of many nonlinear systems can frequently be represented as the sum of a linear function $l(.)$ and a nonlinear function $n(.)$, *i.e.*:

$$
\begin{aligned}
y(k) &= f(y(k-1), y(k-2), \ldots, u(k-1), u(k-2), \ldots) \\
&= l(y(k-1), y(k-2), \ldots, u(k-1), u(k-2)) \\
&\quad + n(y(k-1), y(k-2), \ldots, u(k-1), u(k-2))
\end{aligned} \tag{2.16}
$$

Systems described by Equations (2.7) and (2.8) fall into the category of such nonlinear systems.

Similarly, assuming the input vector of a neural network is $\mathbf{U}$, the scalar output of a network is y_{net}, the nonlinear function vector is $\mathbf{F}$, and the weight matrices of the linear, nonlinear, and hybrid networks are respectively $\mathbf{W}_1$, $\mathbf{W}_2$, $\mathbf{W}_{1a}$, $\mathbf{W}_{1b}$, $\mathbf{W}_{2a}$, and $\mathbf{W}_{2b}$, (see Figures 2.5(a), (b), and (c)), then the output of a linear network is

$$
y_{net} = \mathbf{W}_2 [\mathbf{W}_1 \mathbf{U}] \tag{2.17}
$$

the output of a nonlinear network is

$$y_{net} = \mathbf{W}_2 \, \mathbf{F}[\mathbf{W}_1 \, \mathbf{U}] \qquad\qquad (2.18)$$

and the output of a hybrid network is

$$y_{net} = \mathbf{W}_{2a}[\mathbf{W}_{1a} \, \mathbf{U}] + \mathbf{W}_{2b} \, \mathbf{F}[\mathbf{W}_{1b} \, \mathbf{U}] \qquad\qquad (2.19)$$

According to the previous simulation results, a linear network is able quickly to learn to identify a linear system which can be described by a linear mapping. Therefore, for the systems of Equations (2.7) and (2.9) which have a linear component in their dynamics, a hybrid network learns the systems faster because it also has a linear part. After training has started, the linear part of the network learns the linear part of the system with comparatively high speed. The learning of the nonlinear part of the system is then facilitated by the nonlinear part of the network. For this reason, a hybrid network needs less training. In contrast, a nonlinear network has to learn the linear and nonlinear part with the same nonlinear PEs and this slows down the learning of the network. This explains why the hybrid network performs better than the nonlinear network for the systems of Equations (2.7) and (2.9) when the parameters representing the nonlinearity of the system are within given ranges. When the nonlinearity becomes more significant, the linear part plays a less important role, therefore the hybrid network loses its advantage, because more nonlinear PEs facilitate the learning of the nonlinear mapping. This is why the results show that the hybrid network is not so good when the nonlinear parameters become large. Therefore, the incorporation of a linear part in the structure of a network speeds up the identification of some nonlinear systems with a significant linear part. The more important the linear part is, the faster the hybrid network will be in learning.

2.8 Summary

This chapter has described the use of feedforward neural networks to learn to act as input-output models and state-space models of various linear and nonlinear dynamic systems. This chapter has also discussed hybrid feedforward networks, that is networks composed of both linear and nonlinear activation functions, as tools for identifying nonlinear systems with a significant degree of nonlinearity.

References

Bogoliubov, N.N. and Mitropolsky, Y.A. (1961) *Asymptotic Methods in the Theory of Nonlinear Oscillations*, New York: Gordon and Breach.

Chen, S. and Billings, S.A. (1990), Neural networks for nonlinear dynamic system modelling and identification, *International Journal of Control*, 56(2), 319-346.

Cybenko, G. (1989) Approximation by superpositions of a sigmoidal function, *Mathematics of Control, Signals, and Systems*, 2, 303-314.

Funahashi, K. (1989) On the approximate realization of continuous mappings by neural networks, *Neural Networks*, 2, 183-192.

Harber, R. and Unbehauen, H. (1990) Structure identification of nonlinear dynamic systems - a survey on input / output approaches, *Automatica*, 26(4), 651-677.

Hornik, K. (1991) Approximation capabilities of multilayer feedforward networks, *Neural Networks*, 4, 251-257.

Iserman, R., Lachmann, K.H., and Matko, D. (1992) *Adaptive digital control systems*, London: Prentice Hall.

Narendra, K.S. and Parthasarathy, K. (1990) Identification and control of dynamic systems using neural networks, *IEEE Trans. on Neural Networks*, 1(1), 4-27.

Nguyen, D.H., and Widrow, B. (1990) Neural networks for self-learning control systems, *IEEE Control Systems Magazine*, 18-23.

Ogata, K, (1970) *Modern Control Engineering*, Englewood Cliffs, N.J.: Prentice-Hall, 647.

Pham, D.T. and Liu, X. (1990) State space identification using neural networks, *Engineering Applications of Artificial Intelligence, 3*, 198-203.

Rumelhart, D. and McClelland, J. (1986) *Parallel Distributed Processing: Exploitations in the Micro-Structure of Cognition*, 1 and 2, Cambridge: MIT Press.

Srinivasan, B., Prasad, U.R. and Rao, N.J. (1994) Backpropagation through adjoints for the identification of nonlinear dynamic systems using recurrent neural networks, *IEEE Trans. on Neural Networks*, 5(2), 213-228.

Widrow, B., and Hoff, M. E. Jr. (1960) Adaptive switching circuits, *1960 IRE WEST-CON Conv. Record*, part 4, 96-104.

Yamada, T., and Yabuta, T. (1990) Plant identification using neural control of dynamical systems using neural networks, *IEEE Trans. on Neural networks*, 1(1), 4-27.

Yamada, T. and Yabuta, T. (1990) Plant identification using neural networks, *Japan - USA Symposium on Flexible Automation*, Kyoto, Japan, July 1990, 283-288.

Chapter 3 Dynamic System Identification Using Recurrent Neural Networks

As mentioned in Chapter 1, neural networks can be classified as feedforward networks and recurrent networks. In feedforward networks, the processing elements are connected in such a way that all signals flow in one direction from input units to output units. In recurrent networks there are both feedforward and feedback connections along which signals can propagate in opposite directions.

Feedforward networks have been applied to dynamic system identification with success [Bhat and McAvoy, 1990][Narendra and Parthasarathy, 1990][Yamada and Yabuta, 1990]. Because a feedforward network does not have dynamic memory, the tapped-delay-line method is usually adopted to enable it to represent a dynamic system. The method employs the current and past inputs and outputs of the system to be modelled as the inputs to the network. The next output of the system is used as a teaching signal. The tapped-delay-line method thus turns a temporal modelling problem (learning the dynamic behaviour of the system in the time domain) into a spatial modelling problem (statically mapping the delayed inputs to the next output). However, there are several drawbacks associated with the tapped-delay-line method. One of the drawbacks is slow computation due to the large number of units in the input layer. This is because, if the order of the system to be identified is unknown, it must be over-estimated and hence a large number of units in the input buffer must be given to accommodate large system orders. This usually leads to a large network structure which slows computation (unless a truly parallel computing device is employed). The large number of units in the input layer also makes the identifier highly susceptible to external noise. Another drawback is that the training structure is different from the recall structure and frequently the networks seem to have been trained well with the training structure but have poor performance with the recall structure. For this reason, it is not easy to obtain satisfactory independent simulators of a dynamic system.

Due to their structure, recurrent networks do not suffer from the above drawbacks. Recurrent networks can be classified as fully and partially recurrent. Fully recurrent networks can have arbitrary feedforward and feedback connections, all of which are trainable. In partially recurrent networks, the main network structure is feedforward. The feedforward connections are trainable. The feedback connections are formed through a set of "context" units and are not trainable. The context units memorise some past states of the hidden units, and so the outputs of the networks depend on an aggregate of the previous states and the current input. It is because of this property that partially recurrent networks possess the characteristic of a dynamic memory.

Among the available recurrent networks, the Elman network [Elman, 1990] is one of the simplest types that can be trained using the standard backpropagation learning algorithm [Rumelhart and McClelland, 1986]. This network is originally designed for speech processing applications. This chapter describes the use of the network to identify dynamic systems.

3.1 Basic Elman Network

3.1.1 Structure and Principle of Elman Network

The block diagram of an Elman network is shown in Figure 3.1. From this, it can be seen that in an Elman network, in addition to the input units, hidden units, and output units, there are also context units, as is the case with partially recurrent networks in general. The input and output units interact with the outside environment, while the hidden and context units do not. The input units are only buffer units which pass the signals without changing them. The output units are linear units which sum the signals fed to them. The hidden units can have linear or non-linear activation functions. The context units are used only to memorise the previous activations of the hidden units and can be considered to function as one-step time delays. The feedforward connections (unfilled arrows) are modifiable; the recurrent connections (filled arrows) are fixed. Because the recurrent connections are fixed, the Elman network is only partially recurrent.

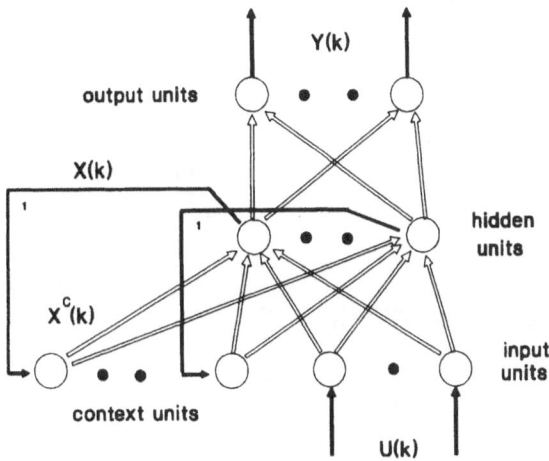

Figure 3.1 An Elman network

At a specific time k, the previous activations of the hidden units (at time $k - 1$) and the current inputs (at time k) are used as inputs to the network. At this stage the network is a feedforward network. These inputs are propagated forward to produce the outputs. The standard back-propagation learning rule [Rumelhart and McClelland, 1986] is then employed to train the network. After this training step, the activations of the hidden units at time k are sent back through the recurrent links to the context units and saved there for the next training step (time $k + 1$). At the beginning of the training, the activations of the hidden units are unknown. Usually, they are set to one half of the maximum range of the value that the hidden units can take. For a sigmoidal activation function, the initial values can be set to 0.5, and for a hyperbolic tangent activation function, they can be set to 0.0.

3.1.2 Analysis of Elman network

In Figure 3.1, the external inputs to the network are represented by $U(k-1)$ and the network outputs, by $Y(k)$. The activations of the hidden units are $X(k)$. The outputs of the context units are represented by $X^c(k)$. From Figure 3.1 it can be seen that the following equations hold

$$\mathbf{X}(k) = \mathbf{F}\{\mathbf{W}^{xc}\mathbf{X}^c(k),\ \mathbf{W}^{xu}\mathbf{U}(k-1)\} \tag{3.1a}$$

$$\mathbf{X}^c(k) = \mathbf{X}(k-1) \tag{3.1b}$$

$$\mathbf{Y}(k) = \mathbf{W}^{yx}\mathbf{X}(k) \tag{3.1c}$$

where $\mathbf{W}^{xc}, \mathbf{W}^{xu}, \mathbf{W}^{yx}$ are weight matrices and $\mathbf{F}$ is a non-linear vector function. In particular, when linear hidden units are adopted and the biases of the hidden and output units are assumed to be zero, Equations (3.1a) and (3.1b) become

$$\mathbf{X}(k) = \mathbf{W}^{xc}\mathbf{X}^c(k)\ +\ \mathbf{W}^{xu}\mathbf{U}(k-1) \tag{3.2a}$$

$$\mathbf{X}^c(k) = \mathbf{X}(k-1) \tag{3.2b}$$

$$\mathbf{Y}(k) = \mathbf{W}^{yx}\mathbf{X}(k) \tag{3.2c}$$

Equations (3.2a) and (3.2c) are standard state-space descriptions of dynamic systems. The order of the models depend on the number of states, which is also the number of hidden units.

When the network is used to model single-input single-output (SISO) systems, only one unit is needed in the input layer and the output layer. For such networks Equations (3.2a) and (3.2c) can be expanded into the following:

$$\begin{aligned}
y(k) = {} & A_1 y(k-1) + A_2 y(k-2) + \ldots + A_n y(k-n) \\
& + B_1 u(k-1) + B_2 u(k-2) + \ldots + B_n u(k-n)
\end{aligned} \tag{3.3}$$

Therefore, theoretically an Elman network is able to model an nth-order dynamic system (where n is the number of units in the hidden layer) if it can be trained to do so. To model the system represented by Equation (3.3), $2n$ input units would be needed if the tapped-delay-line method is used. For an Elman network, the number of input units is one, or $n+1$ if the context units are regarded as input units. Thus an Elman network will be significantly smaller in structure than a feedforward network when n is large.

In training an Elman network, an input string is sent to the network and an output string is used as the teaching signal. The dynamic behaviour is provided by the internal feedback connections. The difference between the tapped-delay-line method and an Elman network can also be considered to mirror that between identification methods based on the input-output and state-space descriptions of a dynamic system. However, in an Elman network the system states are not directly used for training.

3.2 Modified Elman Network

Elman networks with PEs having linear activation functions have been used to identify linear systems. During computer simulations, it has been found that the basic Elman networks can identify first order linear systems easily but have difficulties in identifying linear systems higher than the first-order. Large learning rates cause oscillations or even instabilities to the training process. When suitably small learning rates are adopted so that no oscillations or instabilities occur, training RMS errors are extremely slow to reach an acceptable error level for good recall results. Increasing the number of hidden units makes the achievable RMS error levels smaller. However the number of hidden units cannot be too large because the permissible learning rates become even smaller and the training is even slower. No successful results have been obtained with the basic Elman network even though a large number of tests have been carried out for linear systems higher than the first-order.

Based on the observation that the original Elman network can identify first-order linear systems but not higher-order linear systems, it has been decided to introduce self-connections in the context units of the network in a similar way as described in [Jordan,1986] for other networks, to give these units "inertias ", thereby improving the dynamic memorisation ability of the network. The modification made to the basic Elman network is depicted in Figure 3.2. From Figure 3.2, the output of the j th context unit in the modified Elman network is given by:

$$x_j^c(k) = \alpha x_j^c(k-1) + x_j(k-1) \tag{3.4}$$

where $x_j^c(k)$ and $x_j(k)$ are respectively the outputs of the j th context unit and j th hidden unit and α is the feedback gain of the self-connections. The value of α adopted is the same for all self-connections and is not modified by the training algorithm.

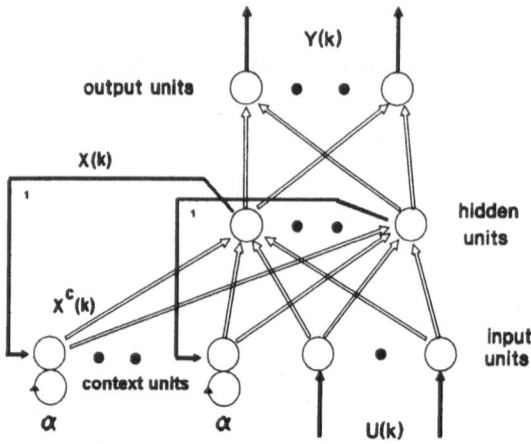

Figure 3.2 The modified Elman network

From Equation (3.4):

$$x_j^c(k) = x_j(k-1) + \alpha x_j(k-2) + \alpha^2 x_j(k-3) + \ldots \tag{3.5}$$

Equation (3.5) shows that the output of the context unit is an integration of the output of the hidden unit that it is connected to. Usually α is between 0 and 1. A value of α nearer to 1 enables the context unit to aggregate more past outputs. Since the order of a dynamic system is related to the number of past outputs on which the present output depends, the introduction of self-feedback in the context units increases the possibility of the Elman network to model higher-order systems.

3.3 Dynamic System Identification

The modified Elman network is proposed to identify high-order linear systems because it is difficult for the basic Elman network to identify them. During simulations it has been found that with different dynamic systems there are different values for the gains of the self-feedback links of the context units that enable the modified Elman network to have the best performance. These values can be found manually by varying the gains from 0 to 1 in steps of 0.1. Obviously this process is inconvenient and time consuming.

Because higher-order linear systems cannot be identified by the basic Elman network but can be identified by the modified Elman network with fixed gains in the context units, to test the performance of the above learning algorithm, only higher-order linear system models are employed. The hidden layers of all networks have linear activation functions.

2nd-order linear system

The 2nd-order linear system model is

$$G(s) = \frac{1 - e^{-Ts}}{s} \frac{\omega}{(s+a)^2 + \omega^2} \tag{3.6}$$

Its discrete form is

$$y(k) = A_1 y(k-1) + A_2 y(k-2) + B_1 u(k-1) + B_2 u(k-2) \tag{3.7}$$

With the sampling period T = 0.1s and the parameters a = 1.0 and $\omega = 2\pi / 2.5$, the coefficients of (3.7) are: A_1 = 1.752821, A_2 = - 0.818731, B_1 = 0.011689, and B_2 = 0.010942.

A linear network with 1 input unit, 6 hidden / context units and 1 output unit is used to identify the system represented by Equation (3.7). A training set of 400 data points is produced by sending a uniformly random bounded sequence $|u(k)| \le (a^2 + \omega^2) / \omega$ (= 2.911160) to the system model with zero initial conditions and recording the output data. After training, the network is tested using a step input signal $u(k) = (a^2 + \omega^2) / \omega$.

A group of simulations are carried out to compare the modified Elman network with α fixed at 0.6 (trained with the standard backpropagation algorithm) and the basic Elman network. The simulations are conducted with all test conditions kept constant (learning rate = momentum = 0.02, seed for the random number generator which produces $u(k)$ = 2, training epochs = 400) except for the initial weights which are generated using a random number generator with a different seed in each case. The example responses of the networks are shown in Figures 3.3(a)-(b).

3nd-order linear system

The second linear system to be identified is a 3rd-order system with one real pole and two complex poles given by:

$$G(s) = \frac{1 - e^{-Ts}}{s} \frac{1}{(s+b)[(s+a)^2 + \omega^2]}$$

(3.8)

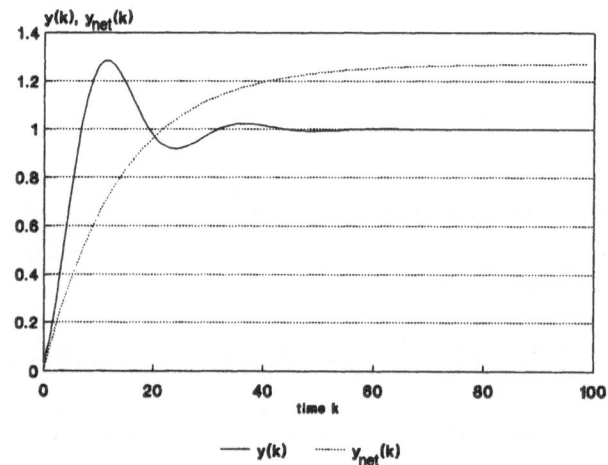

Figure 3.3 (a) Responses of the 2nd order linear system: $\alpha = 0$.

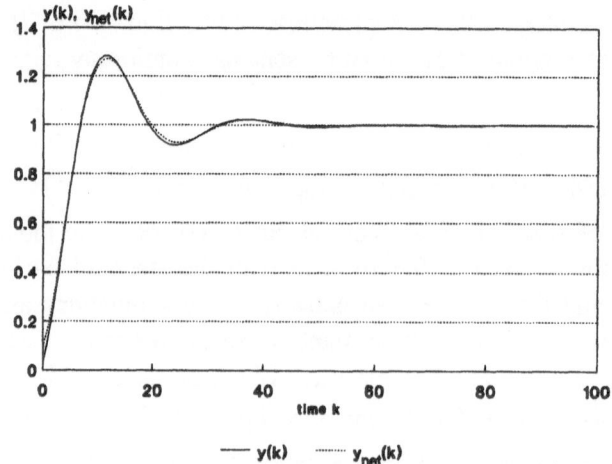

Figure 3.3 (b) Responses of the 2nd order linear system: $\alpha = 0.6$

Its discrete form is

$$y(k) = A_1 y(k-1) + A_2 y(k-2) + A_3 y(k-3)$$
$$+ B_1 u(k-1) + B_2 u(k-2) + B_3 u(k-3) \qquad (3.9)$$

with the sampling period $T = 0.08$s, and the characteristic parameters $a = 1.0$, $b = 2.5$, and $\omega = 2\pi / 2.5$. The coefficients of Equation (3.9) are: $A_1 = 2.627771$, $A_2 = -2.333261$, $A_3 = 0.697676$, $B_1 = 0.017203$, $B_2 = -0.030862$, and $B_3 = 0.014086$. Again a uniformly random bounded sequence $|u(k)| \leq b(a^2 + \omega^2)$ is applied to the system with zero initial conditions and the output of the system is recorded to produce a training data file. The network is tested with a step input $u(k) = b(a^2 + \omega^2)$. A basic Elman network and a modified Elman network with $\alpha = 0.7$ are employed in simulations. The example responses obtained are shown in Figures 3.4(a) and (b).

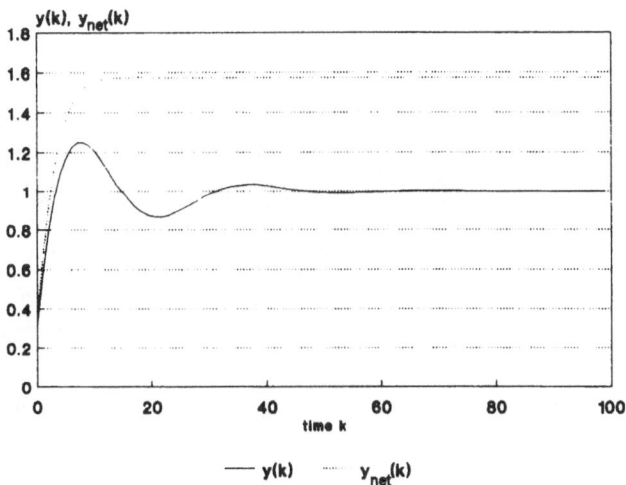

Figure 3.4 (a) Responses of 3rd-order linear system: $\alpha = 0$.

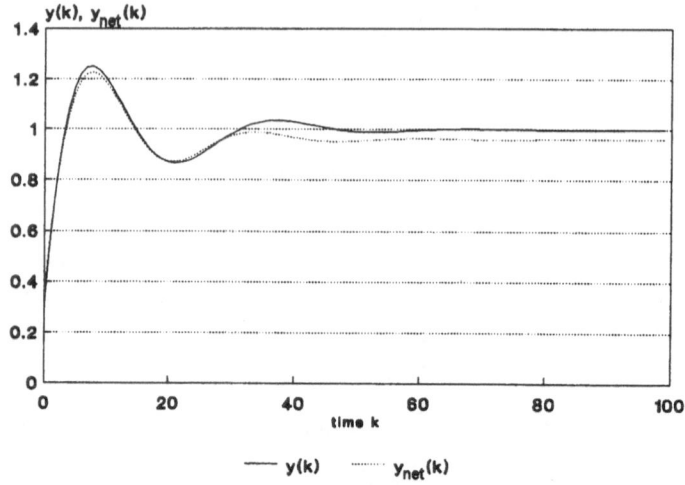

Figure 3.4 (b) Responses of 3rd-order linear system: (b) $\alpha = 0.7$.

3.4 Further Analysis of Elman Networks

3.4.1 Dynamic Backpropagation in Elman Networks

When the weights of the basic Elman network (equations (3.1a-c) are estimated recursively, the feedback $\mathbf{X}^c(k)$ depends on $\mathbf{X}(k-1)$ which can be represented by $\mathbf{W}_{k-1}^{xc}\mathbf{X}^c(k-1) + \mathbf{W}_{k-1}^{xu}\mathbf{U}(k-2)$. $\mathbf{X}^c(k-1)$ depends on $\mathbf{X}(k-2)$ which is equal to $\mathbf{W}_{k-2}^{xc}\mathbf{X}^c(k-2) + \mathbf{W}_{k-2}^{xu}\mathbf{U}(k-3)$. Therefore, $\mathbf{X}^c(k)$ depends on the weights of different previous time instants. When the backpropagation method is applied the dependence of $\mathbf{X}^c(k)$ on the weights should also be taken into account. The backpropagation algorithm so derived is termed a dynamic backpropagation algorithm [Kuan, 1989].

To obtain the dynamic backpropagation algorithm, consider equations (3.2a) to (3.2c). Assume there is only one input unit and one output unit. The training data set is $(u(k), y_d(k))$, $k = 1, 2, ... , N$. When one input-output data pair is presented to the network at time k, the squared error at the network output is defined as:

$$E_k = \frac{1}{2}(y_d(k) - y(k))^2 \tag{3.10}$$

For the whole training data set, the summed squared error is:

$$E = \sum_{k=1}^{N} E_k \tag{3.11}$$

If pattern-based learning is conducted, the weights are modified at each time step k. For $\mathbf{W}^{yx}$

$$\frac{\partial E_k}{\partial \mathbf{W}^{yx}} = -(y_d(k) - y(k))\frac{\partial y(k)}{\partial \mathbf{W}^{yx}}$$

$$= -(y_d(k) - y(k))\mathbf{X}^T(k) \tag{3.12}$$

For $\mathbf{W}^{xu}$ and $\mathbf{W}^{xc}$,

$$\frac{\partial E_k}{\partial \mathbf{W}^{xu}} = -\frac{\partial E_k}{\partial y(k)}\frac{\partial y(k)}{\partial \mathbf{X}(k)}\frac{\partial \mathbf{X}(k)}{\partial \mathbf{W}^{xu}}$$

$$= -(y_d(k) - y(k))\mathbf{W}^{yx}{}^T u(k) \tag{3.14}$$

$$\frac{\partial E_k}{\partial w_i^{xc}} = -\frac{\partial E_k}{\partial y(k)}\frac{\partial y(k)}{\partial x_i(k)}\frac{\partial x_i(k)}{\partial w_i^{xc}}$$

$$= -(y_d(k) - y(k))w_i^{yx}\frac{\partial x_i(k)}{\partial w_i^{xc}} \tag{3.15}$$

In Equation (3.15), $x_i(k)$ is the ith element of $\mathbf{X}(k)$, $\mathbf{w}_i^{xc}$ is the ith row of $\mathbf{W}^{xc}$, w_i^{yx} is the ith element of $\mathbf{W}^{yx}$. As discussed above, the internal feedback $\mathbf{X}^c(k)$ $(=\mathbf{X}(k-1))$ is dependent on $\mathbf{W}^{xc}$. Therefore, from Equation (3.2a),

$$\frac{\partial x_i(k)}{\partial w_i^{xc}} = \mathbf{X}^c{}^T(k) + \mathbf{w}_i^{xc}\frac{\partial \mathbf{X}^c(k)}{\partial w_i^{xc}}$$

$$= \mathbf{X}^T(k-1) + \mathbf{w}_i^{xc}\frac{\partial \mathbf{X}(k-1)}{\partial w_i^{xc}} \tag{3.16}$$

Equation (3.16) shows that there is a dynamic trace of the gradient. This is similar to backpropagation through time [Werbos, 1988]. Because the general expression for weight modification in the gradient descent method is

$$\Delta \mathbf{W} = -\eta \frac{\partial E_k}{\partial \mathbf{W}} \qquad (3.17)$$

the dynamic backpropagation algorithm for the Elman network can be summarised as follows:

network:

$$\mathbf{X}(k) = \mathbf{W}^{xc}\mathbf{X}^c(k) + \mathbf{W}^{xu}u(k-1) \qquad (3.18a)$$
$$\mathbf{X}^c(k) = \mathbf{X}(k-1) \qquad (3.18b)$$
$$y(k) = \mathbf{W}^{yx}\mathbf{X}(k) \qquad (3.18c)$$

algorithm:

$$\Delta \mathbf{W}^{yx} = \eta(y_d(k) - y(k))\mathbf{X}^T(k) \qquad (3.19a)$$
$$\Delta \mathbf{W}^{xu} = \eta(y_d(k) - y(k))\mathbf{W}^{yxT}u(k) \qquad (3.19b)$$
$$\Delta w_i^{xc} = \eta(y_d(k) - y(k))w_i^{yx}\frac{\partial x_i(k)}{\partial w_i^{xc}} \qquad (3.19c)$$
$$\frac{\partial x_i(k)}{\partial w_i^{xc}} = \mathbf{X}^T(k-1) + w_i^{xc}\frac{\partial \mathbf{X}(k-1)}{\partial w_i^{xc}} \qquad (3.19d)$$

If the dependence of $\mathbf{X}(k-1)$ on $\mathbf{W}^{xc}$ is ignored, the above algorithm degenerates to the standard backpropagation algorithm:

$$\Delta \mathbf{W}^{yx} = \eta(y_d(k) - y(k))\mathbf{X}^T(k) \qquad (3.20a)$$
$$\Delta \mathbf{W}^{xu} = \eta(y_d(k) - y(k))\mathbf{W}^{yxT}u(k) \qquad (3.20b)$$
$$\Delta w_i^{xc} = \eta(y_d(k) - y(k))w_i^{yx}\frac{\partial x_i(k)}{\partial w_i^{xc}} \qquad (3.20c)$$
$$\frac{\partial x_i(k)}{\partial w_i^{xc}} = \mathbf{X}^{cT}(k) = \mathbf{X}^T(k-1) \qquad (3.20d)$$

3.4.2 Relation with the Modified Elman network

It has been discovered through simulations that a linear Elman net trained by the standard backpropagation algorithm can only model first-order linear systems [Pham and Liu, 1992]. It can be seen from Equation (3.20d) that the gradient only traces back one time step. However, the gradient in Equation (3.19d) traces back infinitely. According to [Robinson and Fallside, 1987], it can be deduced that an Elman net trained using Equation (3.20a)-(3.20d) can only represent a first-order finite impulse response (hence a first-order dynamic system). However, Equations (3.19a)-(d) can train an Elman net to model an infinite impulse response (thus higher order dynamic systems).

Self-feedback links with fixed gains are introduced to the context units in the modified Elman net to enable the Elman network to represent higher order systems. The modified Elman network of [Pham and Liu, 1992] can be described by the following Equations:

network:

$$\mathbf{X}(k) = \mathbf{W}^{xc}\mathbf{X}^c(k) + \mathbf{W}^{xu}u(k-1) \tag{3.21a}$$

$$\mathbf{X}^c(k) = \mathbf{X}(k-1) + \alpha\mathbf{X}^c(k-1) \tag{3.21b}$$

$$y(k) = \mathbf{W}^{yx}\mathbf{X}(k) \tag{3.21c}$$

algorithm:

$$\Delta\mathbf{W}^{yx} = \eta(y_d(k) - y(k))\mathbf{X}^T(k) \tag{3.22a}$$

$$\Delta\mathbf{W}^{xu} = \eta(y_d(k) - y(k))\mathbf{W}^{yxT}u(k) \tag{3.22b}$$

$$\Delta\mathbf{w}_i^{xc} = \eta(y_d(k) - y(k))w_i^{yx}\frac{\partial x_i(k)}{\partial\mathbf{w}_i^{xc}} \tag{3.22c}$$

$$\frac{\partial x_i(k)}{\partial\mathbf{w}_i^{xc}} = \mathbf{X}^{cT}(k) = \mathbf{X}^T(k-1) \tag{3.22d}$$

Substituting Equation (3.22d) into Equation (3.21b) yields

$$\frac{\partial x_i(k)}{\partial\mathbf{w}_i^{xc}} = \mathbf{X}^T(k-1) + \alpha\frac{\partial x_i(k-1)}{\partial\mathbf{w}_i^{xc}} \tag{3.23}$$

Equation (3.23) is very similar in structure to Equation (3.19d). Although Equation (3.23) does not provide exactly the same search direction as Equation (3.19d) (note $\alpha(\partial x_i(k-1)/\partial\mathbf{w}_i^{xc})$ appears in Equation (3.23), but $\mathbf{w}_i^{xc}(\partial\mathbf{X}(k-1)/\partial\mathbf{w}_i^{xc})$ appears in Equation (3.19d)), it can give an infinite impulse response. This is why the modified Elman net is able to model higher-order systems.

3.5 Summary

This Chapter has discussed the use of partially recurrent Elman networks to identify dynamic systems. The basic Elman network structure as originally proposed by Elman should in principle allow the network to represent systems of any order. However, in practice, the basic Elman network can only model first-order systems if the standard backpropagation is used to train it. Simple modification to the basic structure is suggested in the chapter, which has been shown through simulations to enable the network to represent high order systems. The relationship between the modified network and dynamic backpropagation has been clarified and the enhanced ability of the modified network explained.

References

Bhat, N. and McAvoy, T.J. (1990) Use of neural nets for dynamic modelling and control of chemical process systems, *Computers Chem. Engng*, **14**(4/5), 573-583.

Elman, J.L. (1990) Finding structure in time, *Cognitive Science*, **14**, 179-211.

Jordan, M.I. (1986) Attractor dynamics and parallelism in a connectionist sequential machines, *Proceedings of the 8th Annual Conference of the Cognitive Science Society*, 531-546.

Kuan, C.-M. (1989), *Estimation of Neural Network Models*, PhD thesis, University of California, San Diego.

Narendra, K.S. and Parthasarathy, K. (1990) Identification and control of dynamic systems using neural networks, *IEEE Trans. on Neural Networks*, **1**(1), 4-27.

Pham, D.T and Liu, X. (1992) Dynamic system modelling using partially recurrent neural networks, *Journal of Systems Engineering*, **2**(2), 90-97.

Robinson, A.J. and Fallside, F. (1987) The utility driven dynamic error propagation network, CUED-F-INFENT/TR.1(1987), Engineering Department, Cambridge University, England.

Rumelhart, D. and McClelland, J. (1986) *Parallel distributed processing: exploitations in the micro-structure of cognition*, volume 1 and 2, Cambridge: MIT Press.

Werbos, P.J.: Generalization of backpropagation with application to a recurrent gas market model, *Neural Networks*, vol.1, 339-356, 1988.

Yamada, T. and Yabuta, T. (1990) Plant identification using neural networks, *Japan - USA Symposium on Flexible Automation*, Kyoto, Japan, July 1990, 283-288.

Chapter 4 Modelling and Prediction Using GMDH Networks

Chapters 2 and 3 have shown that neural networks can be employed to identify dynamic systems. The main advantages of neural networks over conventional identification methods include simplicity of implementation and good approximation properties [Warwick et al, 1992]. In feedforward network based identification schemes, neural networks are used to represent the implied static mapping between the available input and output data. The network structures (number of layers and number of units in each layer) are predefined and remain unchanged both during and after training. Successful identification is often dependent on proper pre-estimation of the network structure.

GMDH (Group Method of Data Handling) neural networks [Ivakhnenko, 1971; Barron, 1975; Shrier et al, 1987; Hecht-Nielsen, 1990] are layered networks with a structure which is configured through training. The PEs in these networks usually have two inputs. The output of each PE is normally a quadratic combination of its two inputs. The activation functions of the PEs can be considered polynomials of the second degree [Ivakhnenko, 1971; Shrier et al, 1987]. During training, the layer number in a GMDH network increases. Each time a new layer is added, a number of PEs are created. Poorly-performing PEs are discarded and PEs that perform well are preserved to build new layers. In this sense the PE number in a layer is also adaptable. Another advantage of GMDH neural networks is that they do not suffer from the problem of overfitting the training data. In an ordinary fixed-structure network, this problem causes the network to respond poorly to data that it has not been trained on, although it might perform very well with the training data. To avoid overfitting, the GMDH technique prescribes dividing the available experimental input-output data into two sets. One set is used for training and is called the training data set. The other is the selection data set which is used for selecting good PEs.

As stated above, the input-output relation for a PE in a GMDH network can be described by a second degree polynomial. In general the coefficients of the polynomials in a GMDH network are found using linear regression [Hecht-Nielsen, 1990]. The disadvantage of a GMDH scheme based on linear regression is that it involves a large number of high dimensional matrix calculations. In this chapter, a GMDH network is represented as an assembly of N-Adalines [Widrow and Lehr, 1990], that is Adalines [Widrow and Stearns, 1985] having second-degree non-linear preprocessors the weights of which correspond to the polynomial coefficients of the original GMDH network. It is proposed that the Widrow-Hoff learning rule [Widrow and Lehr, 1990] is employed instead of linear regression to train such a network. Therefore, the problem of obtaining the polynomial coefficients has been turned into one of executing a well proven recursive learning procedure. To demonstrate the good performance of the proposed GMDH network paradigm, the results of two example applications will be presented. The first application is the modelling of a non-linear dynamic system based upon simulated input-output data. The second application is the prediction of the level of a river using real level measurements taken over a period of time.

4.1 N-Adaline Networks and Widrow-Hoff Learning

An N-Adaline unit is shown in Figure 4.1. From this, it can be deduced that the output y of the unit is given by:

$$y = w_0 + w_1 x_1 + w_2 x_1^2 + w_3 x_1 x_2 + w_4 x_2^2 + w_5 x_2 \qquad (4.1)$$

where w_i ($i = 0, 1, ..., 5$) are the weights of the unit and x_1 and x_2 are the inputs to the unit. Thus, y is a linear combination of the weights and a polynomial of degree 2 in terms of the input variables. If a number of N-Adalines units are arranged in a layered form, a polynomial network is obtained which can approximate complex mappings of higher degrees.

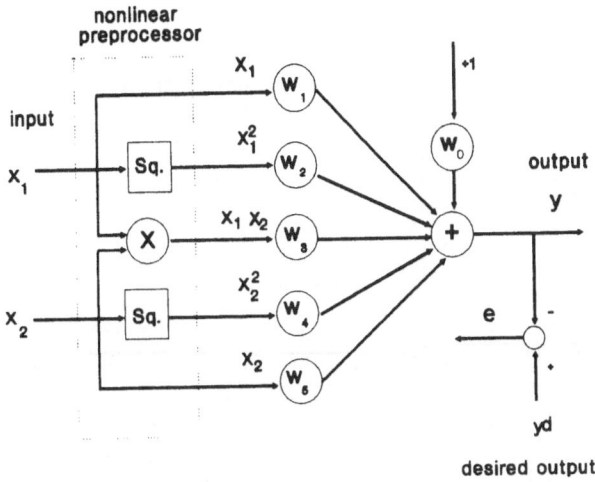

Figure 4.1 An Adaline with nonlinear preprocessor

Let $\mathbf{W} = [w_0\ w_1\ w_2\ w_3\ w_4\ w_5]^T$ and $\mathbf{X} = [1\ x_1\ x_1^2\ x_1x_2\ x_2^2\ x_2]^T$. The Widrow - Hoff delta rule adopted for training $\mathbf{W}$ is as follows [Widrow and Lehr, 1990]:

$$\mathbf{W}_{k+1} = \mathbf{W}_k + \alpha \frac{\mathbf{X}_k}{|\mathbf{X}_k|^2}(y_d^k - \mathbf{W}_k^T\mathbf{X}_k) \tag{4.2}$$

where y_d^k is the desired output at time k and α is the learning rate. The practical range of α is (0.1, 1). The application of Equation (4.2) causes $\mathbf{W}$ to be modified so as to reduce the difference between the desired and actual outputs.

4.2 GMDH Network Based on N-Adalines

An N-Adaline based GMDH network with 4 external inputs is depicted in Figure 4.2. This is a layered feedforward network which can represent a many-to-one continuous mapping. If a mapping to multiple outputs is required, several networks can be employed, one for each output.

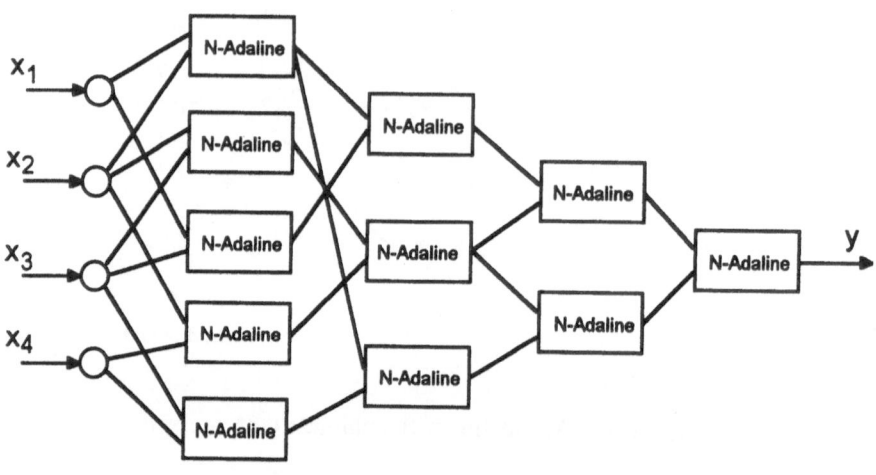

Figure 4.2 A trained GMDH network

Based on the number of available external inputs, the network grows its first layer of PEs (N-Adalines). Each PE in this layer corresponds to a different pair of inputs. The Widrow-Hoff learning rule is applied to train the PEs using the training data set. Every PE in this layer has the desired output as its target and is expected to produce the correct output. When the batch mean square error (BMSE - mean square errors summed over all the desired outputs in the training data set) of a PE reaches its minimum, its weights are frozen, i.e. the weight modification process is halted for that PE. When weight modification ceases for all PEs in the layer, the training of this layer ends. The selection data set is then employed to select the trained PEs for the next stage. Those PEs whose BMSEs for the selection data are below a given threshold are kept as the "post-selection" PEs of this layer and their outputs are taken as the inputs to the following layer if one is created. The smallest BMSE achieved among all the PEs during the selection process is recorded and used as a criterion for stopping the training of the whole network. Assuming that the smallest BMSE for the current layer is less than that recorded for the previous layer, a new layer is generated, the size of which is determined by the number of PEs just selected. The training and selection procedure is then repeated for the new layer. Existing trained layers are untouched by that procedure. Training and selection are carried out until the smallest BMSE during selection for a new layer is larger than that for the last layer or the new layer has only one PE. If the procedure is stopped because of an increase in the smallest BMSE, the layer before the last is used as the output layer and its best PE is taken as the output unit of the whole network. If the new layer has only one N-Adaline but its BMSE during selection is still smaller than that recorded for the previous layer, then that N-Adaline is adopted as the output unit. To obtain the final structure of the network, all PEs not directly or indirectly linked to the output unit are trimmed off, leaving only those which are relevant to the computation of the output value.

The output of a GMDH network having m layers can be expressed as a (2^m)th degree polynomial. Thus an m-layered GMDH network can be expected to be able to represent a non-linear mapping of (2^m)th degree in nonlinearity.

The method used investigation to train GMDH networks for modelling and prediction is summarised as follows:

(1) Preprocess data. This is to normalise the data and eliminate stationary DC components which are unhelpful in dynamic modelling [Landau, 1990]. As a common practice with the GMDH technique [Yoshimura et al, 1985], the available input-output data are processed before training using the following equations:

$$u_i^* = \frac{u_i - \overline{u}}{\sigma_u} \tag{4.3}$$

$$y_i^* = \frac{y_i - \overline{y}}{\sigma_y} \tag{4.4}$$

where u_i, y_i are the i th input and output experimental data pair, $\overline{u}$ and $\overline{y}$ are the means of the u_i's and y_i's, and σ_u and σ_y are the standard deviations of the u_i's and y_i's;

(2) Decide the external inputs to the network. For a modelling application, m past inputs $u(k-1)$, $u(k-2)$, ... , $u(k-m)$ and n past outputs $y(k-1)$, $y(k-2)$, ... , $y(k-n)$ are used. For a prediction application, only n past outputs are used. If necessary, m and n can be determined by

calculating correlation coefficients for the input output data [Duffy and Franklin, 1975];

(3)   Separate the experimental data set into a training data set and a selection data set;

(4)   Create a layer of N-Adalines based on the number of inputs. Every pair of inputs produces an N-Adaline. Provisionally C_i^2 N-Adalines are created if there are i inputs;

(5)   Initialise the weights of the provisional PEs (N-Adalines) to zero;

(6)   Use the training data set to train all the PEs in the created layer. The training proceeds as follows. At time k, apply $y(k-l)$, $l = 1, 2, ... , n$ (for a prediction task) or $y(k-l)$, $l = 1, 2, ... , n$ and $u(k-l)$, $l = 1, 2, ... , m$ (for a modelling task) to the network. Take $y(k)$ as the desired output of all the PEs. Calculate the output errors of the PEs and modify their weights once. One batch of training is said to have been finished when the whole training data set has been presented to the network. The squared errors of a PE before weight modification are summed over the batch to obtain the BMSE (prediction error) for that PE. If the BMSE for this PE is smaller than that for the previous batch, subject it to a new batch of training. Otherwise stop training it. The training of the present layer ends when the training for all the PEs has ceased;

(7)   Input the selection data set to the network and obtain the BMSEs of all the PEs in the layer just trained and the ratio of each BMSE with respect to the smallest BMSE. Based on the ratios obtained, assign a threshold. Keep those PEs whose ratios are below the threshold and use them as the "post-selection" PEs for the present layer;

(8)   If the smallest BMSE in the layer just selected is larger than that of the previous layer or the present layer has only one unit (assuming the network already has two layers or more), stop the training process. If training is stopped due to the BMSE increasing, use the previous layer as the output layer of the trained network and trim the network. If training is stopped because there is only one unit in the present layer and its BMSE is smaller than the BMSE for the previous layer, use the present layer as the output layer and also trim the network. Otherwise based on the "post-selection" PEs, create a new layer and return to step (5);

(9)   Test the performance of the trained network with the evaluation data set. The latter could be a combination of the training and selection sets, or a completely new data set. As only the training data set is used in the determination of network weights, if the evaluation data set consists of both the training and selection sets, the generalisation ability of the network can

be tested. If only new data are employed for evaluating performance, a larger generalisation range is tested.

4.3 Applications

This section presents the results of using the N-Adaline network in two example applications. The first application is the identification of a non-linear plant with the following input-output model [Narendra and Parthasarathy, 1990]:

$$y(k) = \frac{y(k-1)}{1 + y^2(k-1)} + u^3(k-1) \tag{4.5}$$

where the input variable $u(k)$ is a random number uniformly distributed in the interval (-2.0, 2.0), and the output variable $y(k)$ has initial state $y(0)$ = 0.

Fifty data points are produced using Equation (4.5). The first 20 data are reserved for training, the second 20 data are used for selection, and all the 50 data are employed for evaluation. Networks with two input units (for $u(k-1)$ and $y(k-1)$), four input units (for $u(k-1)$, $u(k-2)$, $y(k-1)$, $y(k-2)$) and six input units (for $u(k-1)$, $u(k-2)$, $u(k-3)$, $y(k-1)$, $y(k-2)$, $y(k-3)$) are employed respectively in three different simulations. The networks obtained after training are shown in Figure 4.3. The responses of the trained nets to the whole set of 50 data are shown in Figure 4.4.

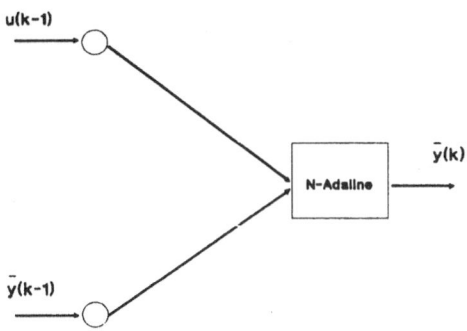

Figure 4.3 (a) GMDH model for nonlinear system: 2 inputs.

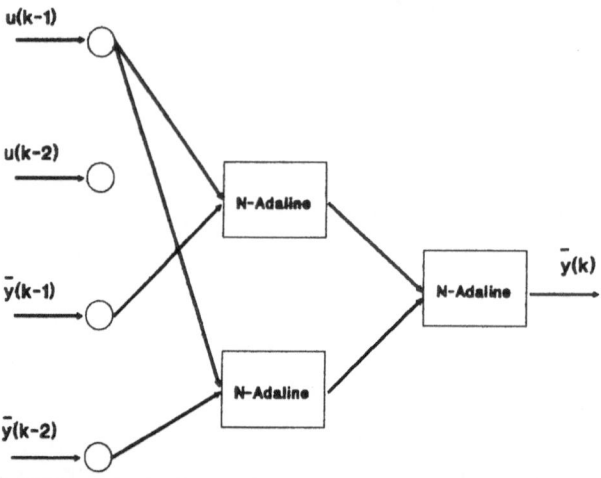

Figure 4.3 (b) GMDH model for nonlinear system: 4 inputs.

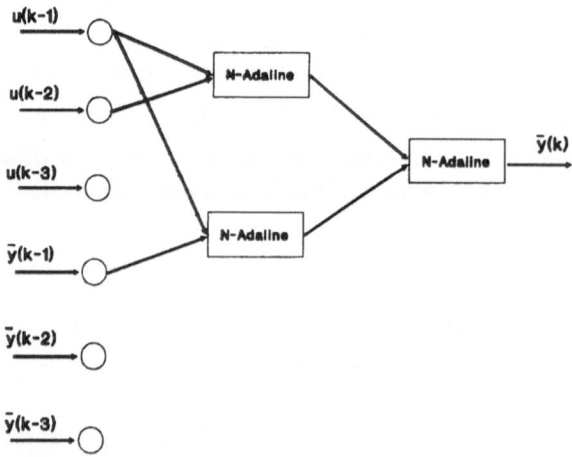

Figure 4.3 (c) GMDH model for nonlinear system: 6 inputs.

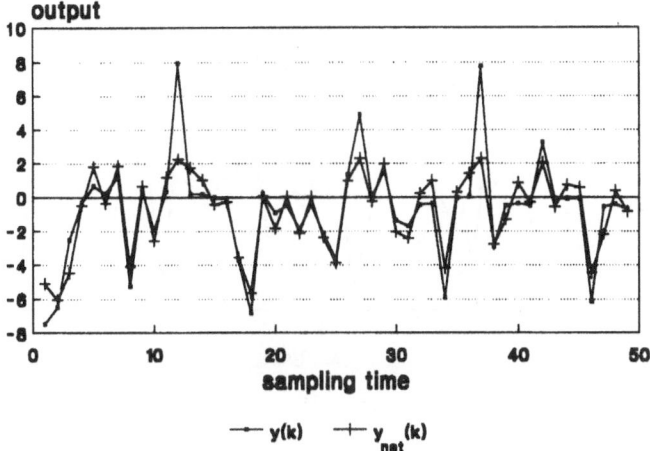

Figure 4.4 (a) GMDH Modelling results: 2 inputs.

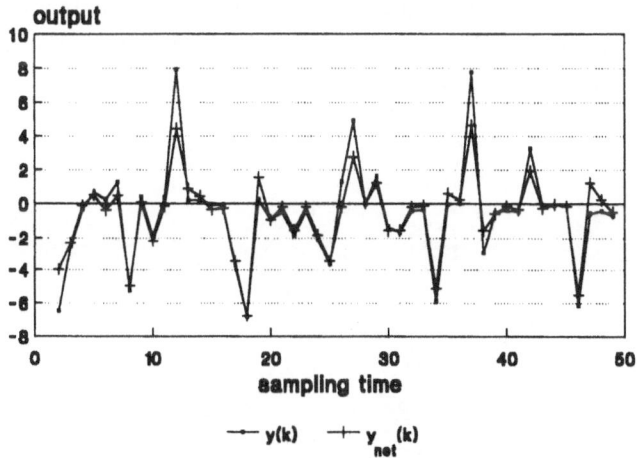

Figure 4.4 (b) GMDH Modelling results: 4 inputs.

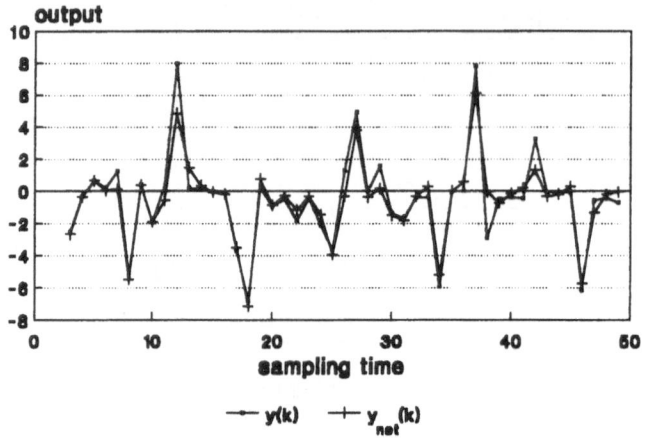

Figure 4.4 (c) GMDH Modelling results: 6 inputs.

Simulations are also conducted to compare the performance of the proposed GMDH network with that of the multilayer perceptron (MLP) network [Rumelhart, D. and McClelland, 1986]. Six hidden units are employed for the MLP network. Figure 4.5 depicts the MLP networks employed and Figure 4.6 shows the results obtained.

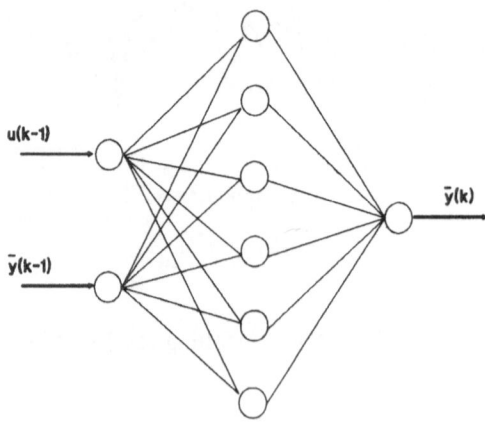

Figure 4.5 (a) MLP modelling of nonlinear plant: 2 inputs.

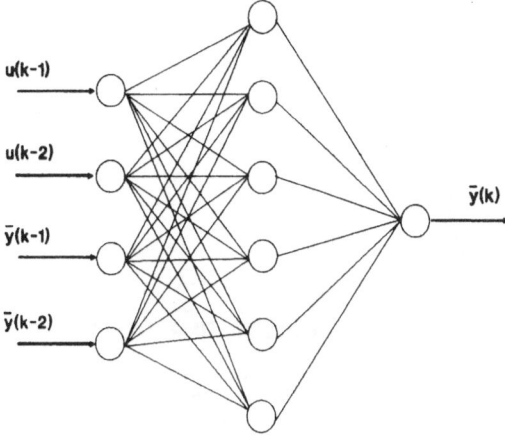

Figure 4.5 (b) MLP modelling of nonlinear plant: 4 inputs.

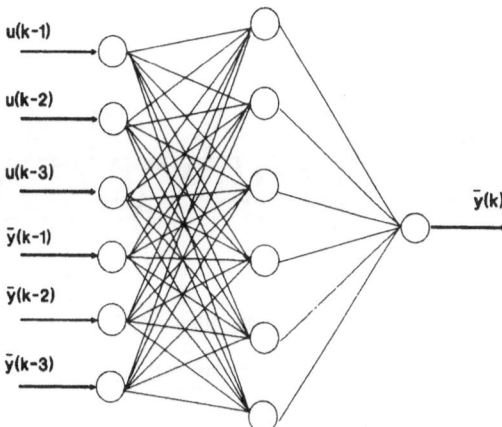

Figure 4.5 (c) MLP modelling of nonlinear plant: 6 inputs.

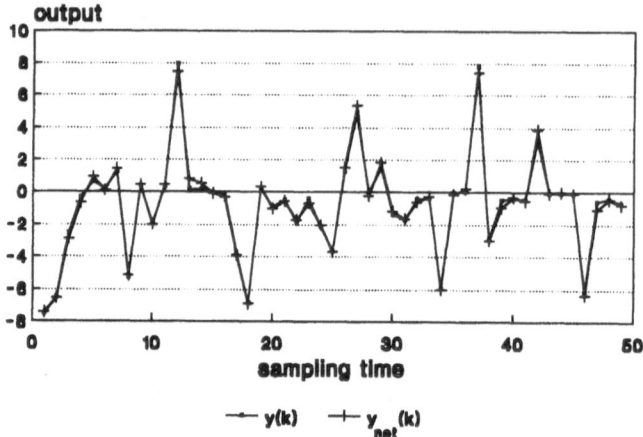

Figure 4.6 (a) MLP modelling results: 2 inputs.

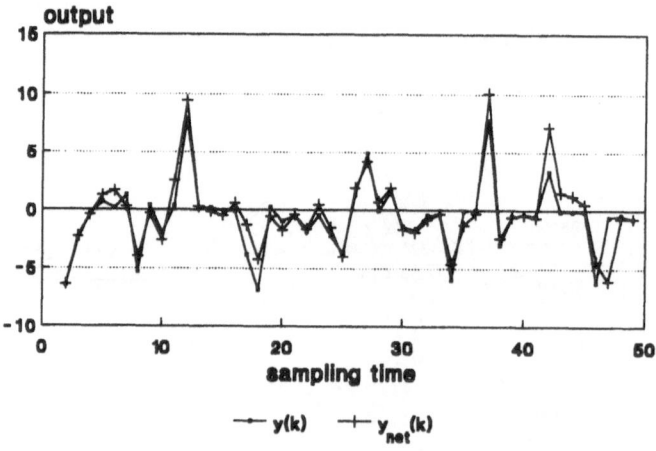

Figure 4.6 (b) MLP modelling results: 4 inputs.

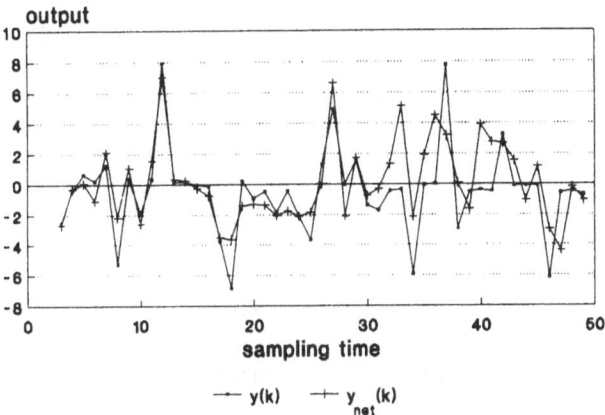

Figure 4.6 (c) MLP modelling results: 6 inputs.

The second example application is the prediction of the level of a river. Table 4.1 lists the annual highest and lowest levels of the Amazon at Iquitos from 1962 to 1978.

This data were originally collected for an investigation into the correlation between deforestation and the increasingly frequent flooding of that river [Weisberg, 1985; Gentry and Lopez-Parodi, 1980]. In the simulation of high water level prediction using the GMDH network, 4 input units are adopted, the first 11 data are used for training and selection, and all the 17 data for evaluation. In the simulation of low water level prediction using the GMDH network, 4 input units are employed, the first 14 data are used for training, 15 data for selection, and 17 data for evaluation. The trained networks and test results obtained are shown in Figures 4.7-4.8. To obtain comparable results for MLP networks and GMDH networks, the MLPs for high and low water level predictions both have 4 input units and used 17 data for evaluation. However, in the simulation of high level prediction, 11 data are used for training and in the low level prediction simulation, 15 training data are employed. The results obtained for the MLP networks are shown in Figure 4.9.

Table 4.1 Amazon river data

Year	High (m)	Low (m)
1962	25.82	18.24
1963	25.35	16.50
1964	24.29	20.26
1965	24.05	20.97
1966	24.89	19.43
1967	25.35	19.31
1968	25.23	20.85
1969	25.06	19.54
1970	27.13	20.49
1971	27.36	21.91
1972	26.65	22.51
1973	27.13	18.81
1974	27.49	19.42
1975	27.08	19.10
1976	27.51	18.80
1977	27.54	18.80
1978	26.21	17.57

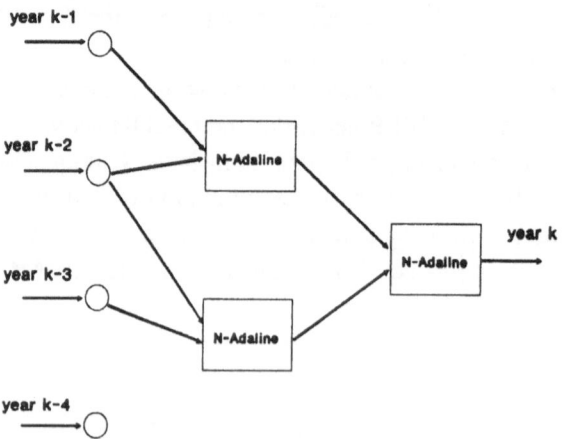

Figure 4.7 (a) GMDH prediction of high water levels: predictor.

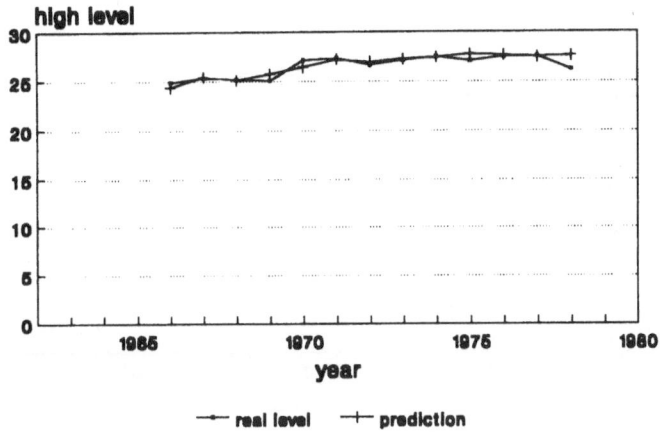

Figure 4.7 (b) GMDH prediction of high water levels: results.

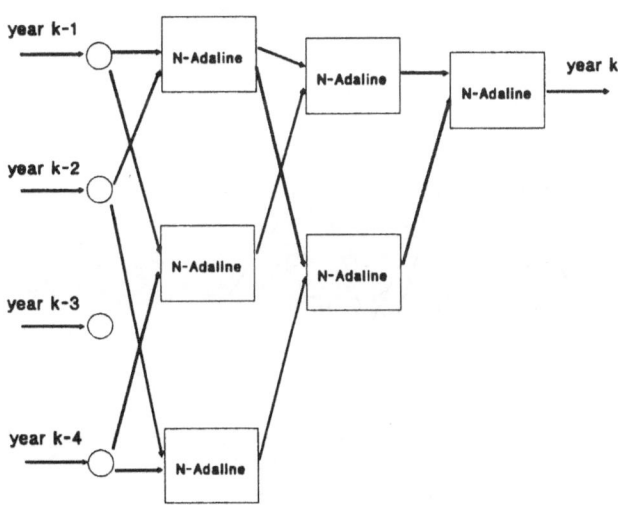

Figure 4.8 (a) GMDH prediction of low water levels: predictor.

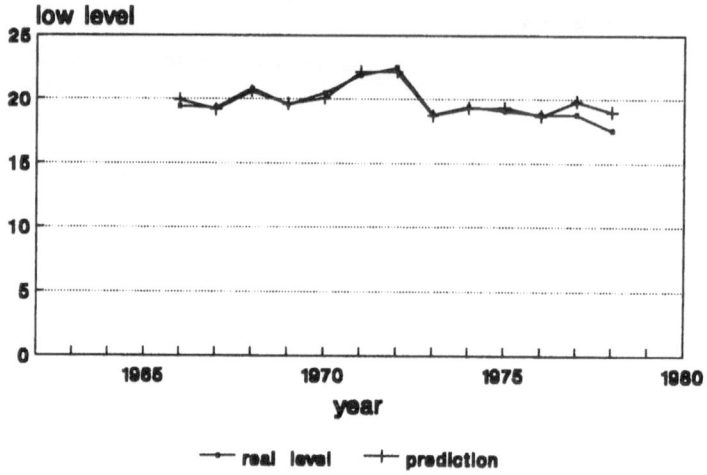

Figure 4.8 (b) GMDH prediction of low water levels: results.

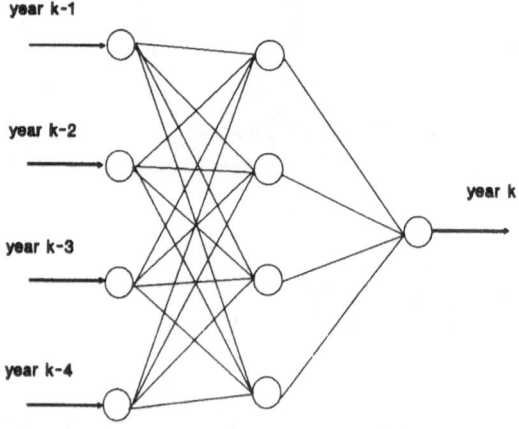

Figure 4.9 (a) MLP prediction of water levels:
predictor for high and low water levels.

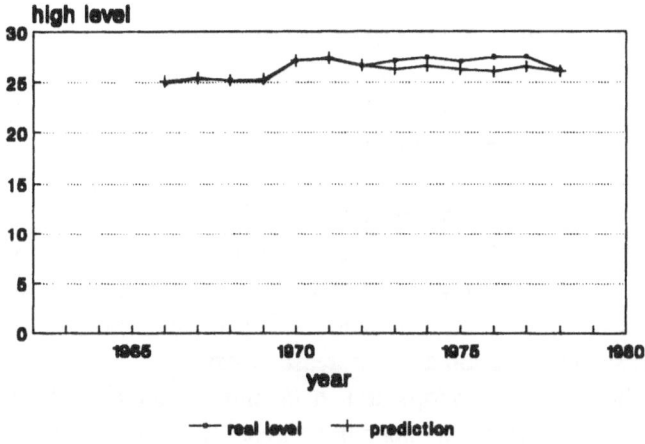

Figure 4.9 (b) MLP prediction of water levels: results for high levels.

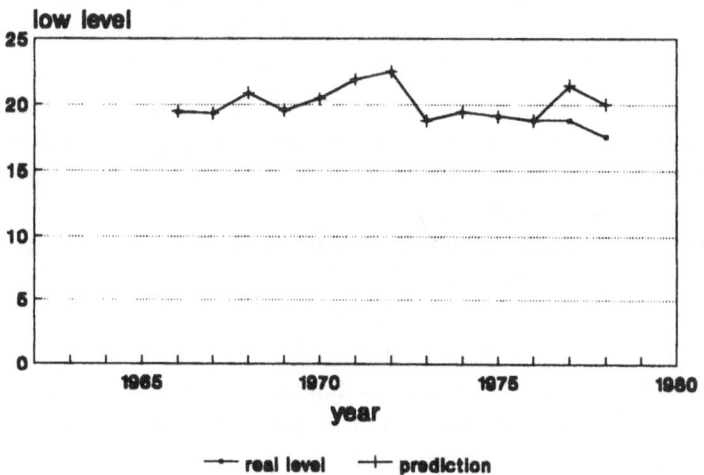

Figure 4.9 (c) MLP prediction of water levels: results for low levels.

4.4 Discussion

The general principle of a GMDH network is to fit a polynomial to a non-linear mapping. The fitting starts with a polynomial of the second degree. If that is not adequate, the degree of the polynomial is increased by 2 and the fitting attempted again. The whole process continues until the best possible result is achieved. The growth and selection strategies of the GMDH scheme produce a network with a flexible structure which always grows in such a direction as to decrease errors.

By splitting the experimental data into training and selection sets, overfitting or pure memorising can be avoided. This is different from other neural network training techniques which usually try to fit the network to one data set. Consequently, a network may learn the training data well but perform poorly with new data. The main issue with the GMDH method is how to split the experimental data set, especially when it is small, as in the water level prediction example.

Due to the use of N-Adalines and Widrow-Hoff learning, the largest matrix encountered in GMDH training is the weight vector the dimension of which is only 6×1. This can be contrasted with the linear regression method described in [Hecht-Nielsen, R. 1990] where the matrix dimensions increase proportionally with the number of training data and a large amount of calculations must be carried out during training. Further discussions of the differences between linear regression and Widrow-Hoff learning can be found in [Kosko, 1992].

From the modelling results it can be seen that the performance of the GMDH network is not satisfactory when it only has two input units. This is because the network can only grow one layer which has just one PE. From this PE a 2nd-degree polynomial is obtained which is not sufficient to approximate the non-linear transfer function of the given plant. When four or six inputs are adopted for the GMDH network, the results are satisfactory. It is interesting to note that the six-input network has results comparable to those given by the 4-input network although according to Equation (4.5) six inputs should be associated with larger errors (the system is 1st-order and requires only $u(k-1)$ and $y(k-1)$ as inputs except for the representation problem mentioned earlier). Due to its correlation property, the GMDH algorithm automatically discards unnecessary inputs during the training, selection, and final trimming processes. In contrast, the performance of the MLP networks is best when there are 2 input units and worsens with the increase of model estimation errors (in networks with 4 to

6 input units). During simulations, it has been found that when 6 input units are adopted, although small training errors can be achieved, large errors are always incurred when the network is used as a model, especially with new input data, as indicated in Figure 4.6(c).

Figures 4.7(b) and 4.9(b) show that in the high water level prediction case, even though only 11 data are used for training and selection, the GMDH network produces reasonable predictions for the period from 1966 to 1978. In particular, the rising trends in high water levels are well predicted. The MLP network gives predictions which are even better than those by the GMDH networks from 1966 to 1972, but performs poorly with predictions from 1973 to 1978. As for the low water level predictions (Figures 4.8(b) and 4.9(c)), the GMDH network produces better predictions for years 1977 and 1978 than the MLP network which in fact has more training data. All these results suggest that the GMDH network has a better ability to generalise.

4.5 Summary

This chapter has discussed the use of Group-Method-of-Data-Handling (GMDH) networks in modelling and prediction tasks. GMDH networks have a structure that grows to fit the particular task being undertaken. This enables them to perform better than networks with fixed structures in the situations discussed in the chapter.

References

Barron, R.L (1975) Learning networks improve computer-aided prediction and control, *Computer Design*, **14**(8), 65-70.

Duffy, J.J. and Franklin, M.A. (1975) A learning identification algorithm and its application to an environment system, *IEEE Trans. Systems, Man, and Cybernetics,* **5**(2), 226-240.

Gentry, A.H. and Lopez-Parodi, J. (1980) Deforestation and increased flooding of the upper Amazon, *Science*, **210**(19), 1354-1356.

Hecht-Nielsen, R. (1990) *Neurocomputing*, Addison-Wesley Publishing Company.

Ivakhnenko, A.G. (1971) Polynomial theory of complex systems, *IEEE Trans. Systems, Man, and Cybernetics*, 12, 364-378.
Kosko, B. (1992) *Neural Networks and Fuzzy system*, Englewood Cliffs, NJ: Prentice-Hall.

Landau, I.D. (1990) *System identification and control design*, Englewood Cliffs, NJ: Prentice Hall.

Shrier, S., Barron, R.L., Gilstrap, L.O. (1987) Polynomial and neural networks: analogies and engineering applications, *Proceedings of IEEE International Conference on Neural Networks*, II, 431-439.

Warwick, K., Irwin, G.W., and Hunt, K.T. (eds) (1992) *Neural networks for control and systems*, Peter Peregrinus, London.

Weisberg, S. (1985) *Applied Linear Regression*, Wiley, New York, 31-32.

Widrow, B. and Lehr, M.A. (1990) 30 years of adaptive neural networks: perceptron, Madaline, and backpropagation, *Proceedings of IEEE*, 78, 1415-1442.

Widrow, B. and Stearns, S.D. (1985) *Adaptive Signal Processing*, Prentice-Hall, Englewood Cliffs, NJ.

Yoshimura, T., Deepak, D., and Takagi, H. (1985) Track / vehicle system identification by a revised group method of data handling (GMDH), *International Journal of Systems Science*, 16(1), 131-144.

Chapter 5 Financial Prediction Using Neural Networks

This chapter discusses the application of neural networks to stock market and exchange rate prediction. The first section illustrates the use of the multilayer perceptron network for stock market prediction. The remainder of the chapter describes experiments with the multilayer perceptron network, the GMDH network, and the Elman network in exchange rate prediction.

5.1 Stock Market Prediction

The stock market is a free market economic system whose dynamics basically depends on the law of supply and demand. Various economic activities and psychological factors affect the rise and fall of stock prices. Due to stochastic human-factors and the nonlinear, multivariable, and temporal nature of stock price transitions, it is difficult to conduct prediction using conventional techniques.

Neural networks have been found useful in stock price prediction [White, 1988; Kamijo and Tanigawa, 1990; Kimoto and Asakawa, 1990; Lee and Park, 1992]. Both feedforward and recurrent neural networks have been investigated and good results have been obtained. In the following section the work of Kimoto and Asakawa [Kimoto and Asakawa, 1990] is described as an example of the use of feedforward neural networks in this area.

5.1.1 System Overview

The prediction system is shown in Figure 5.1.

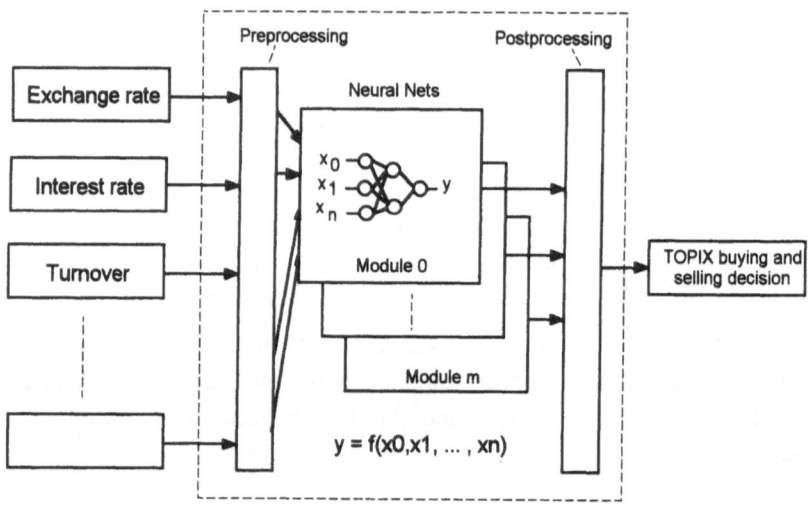

Figure 5.1 Basic structure of Kimoto's and Asakawa's stock price prediction system (adapted from [Kimoto and Asakawa, 1990])

The system consists of several neural network modules. These modules are all used to learn the relationships between different technical and economical indices and the decision to buy and sell stocks. Each module is a three-layer perceptron and has its own training data set. The outputs of the networks are postprocessed by averaging to give the output of the system. The inputs to the networks are technical and economic indices. The output of the system is the decision to buy and sell. In Figure 5.1, TOPIX stands for "Tokyo Stock Exchange Price Indices".

A high-speed backpropagation learning algorithm, called supplementary learning, has been used to train the networks. Before learning starts, tolerances are defined for the output units. During learning, the weights are updated only when the output errors exceed the tolerances. The learning data for which the output errors do not exceed the tolerances are eliminated from the training data sets. By doing this, the amount of calculation for error backpropagation is reduced as learning proceeds. Supplementary learning also changes the learning parameter according to the amount of training data. The learning parameter is defined as a constant divided by the number of training data. As learning proceeds, changes in the number of training data adjust the learning parameter.

The input data to each network are the moving averages of the weekly averaged data of indices such as turnover, interest rate, foreign exchange rate, New York Dow-Jones average, etc. These data are first preprocessed by log or other functions and then by normalisation functions. The teaching data for buy/sell decisions is the weighted sum of weekly returns. The input data and teaching data are all scaled into the [0, 1] range. In prediction applications, an output value of 0.5 or more means "buy" whereas an output value less than 0.5 indicates "sell". As mentioned above, the output value is obtained by averaging the predictions of all network modules.

5.1.2 Prediction Simulation

A simulation has been conducted for data collected over 33 months from January 1987 to September 1989. The results obtained are shown in Figure 5.2.

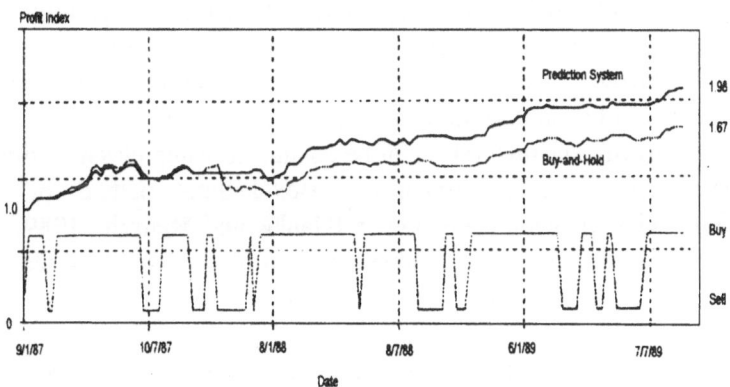

Figure 5.2 Performance of the prediction system (adapted from [Kimoto and Asakawa, 1990])

In Figure 5.2, the dotted line indicates the profit index achieved with the "buy-and-hold" strategy and the solid line, the profit index obtained using the prediction system. The profit index for January 1987 is taken to be 1.00. The profit index achieved with the "buy-and-hold" strategy is 1.67 in

September 1989. Adopting the prediction system produces a profit index of 1.98 which is better than for the "buy-and-hold" strategy.

5.2 Currency Exchange Rate Prediction

Currency exchange rates are an important economic index in the international monetary market. Forecasts of exchange rates are useful to governments and companies in making decisions on investment and trading. Different forecasting approaches have been proposed. Based on analyses of the relationships between exchange rates and wholesale and retail prices, inflation rates, interest rates, and import/export prices etc. a comprehensive forecasting method is introduced in [Lesseps and Morrel, 1977]. However, some researchers [Refenes et al, 1993] believe that all changes in economic policies and other indices are ultimately reflected in the currency exchange rates and the necessary knowledge for predicting the behaviour of an economy is embodied in the currency exchange rates. Moreover, exchange rates can be predicted by extracting regularities from observations of past exchange rates. The work described in this section assumes that the time series of currency exchange rates contain all relevant information and predictions can be made from the historical data in the series.

Predictions of exchange rates can be conducted using classical time series methods such as autoregression (the Box-Jenkins method, or ARIMA method) and exponential smoothing [Hanke and Reitsch, 1989; Brown, 1963] and using multilayer perceptron neural networks [Refenes et al, 1993]. The work of [Refenes et al, 1993] is a pioneering study in applying neural networks to exchange rate prediction, the results of which indicate that multilayer perceptron networks are able to perform well because they are suitable for non-linear prediction. The ARIMA model for regression which is a linear model and the exponential smoothing method which only carries out piecewise linear approximation do not yield as good a performance. The neural network used in [Refenes et al, 1993] is a fixed-structure network.

This section describes investigation into the feasibility of the GMDH network - for time-series prediction. Simulation studies have also been carried out to compare the performance of GMDH networks with those of other feedforward neural networks and of the Elman network.

5.2.1 Prediction Based on Neural Networks

Feedforward (including GMDH and multilayer perceptron or MLP) networks and recurrent neural networks can be employed for exchange rate prediction. The structure of a prediction system based on feedforward neural networks is shown in Figures 5.3(a)-(b).

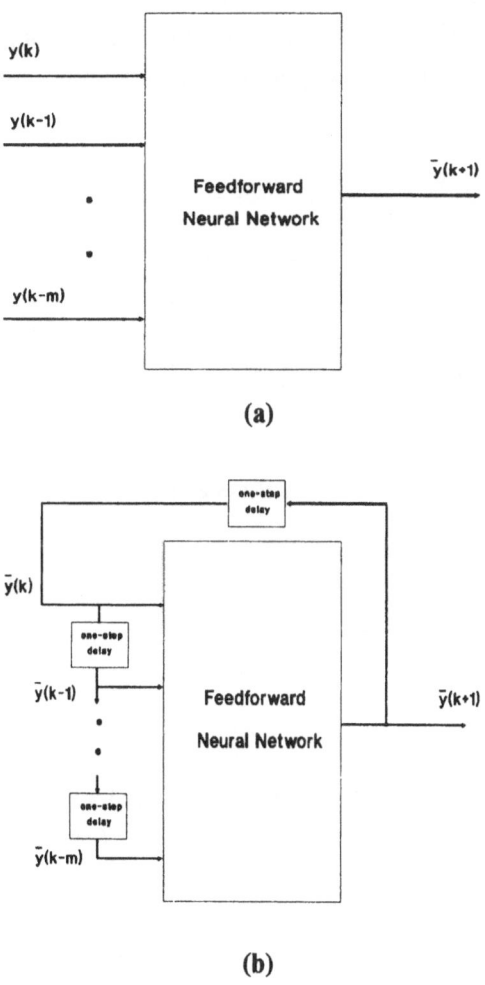

Figure 5.3 Prediction using feedforward neural networks:
(a) One-step; **(b)** Multi-step

For one-step prediction, the inputs to the network are historical rates obtained from real measurements. The network is expected to predict a new rate. When a feedforward network is used for multi-step prediction, its output is fed back to its input layer. Several past outputs so fedback are employed as inputs to the network to provide a prediction for the next output. In the recurrent network case, only the real present rate is required to predict the next rate when the network is used for one-step prediction (Figure 5.4(a)). In multi-step prediction using a recurrent network, the output of the recurrent network is fed back to its input layer and the network recursively predicts new outputs based on only an immediately preceding output (Figure 5.4(b)). In both cases, one-step prediction provides short-term forecasts and multi-step prediction provides long-term forecasts.

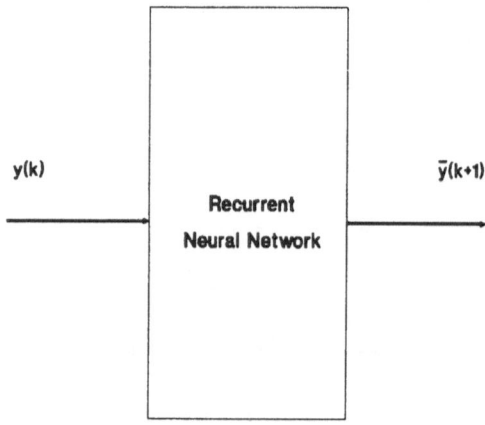

Figure 5.4 (a) Prediction using recurrent neural networks: One-step.

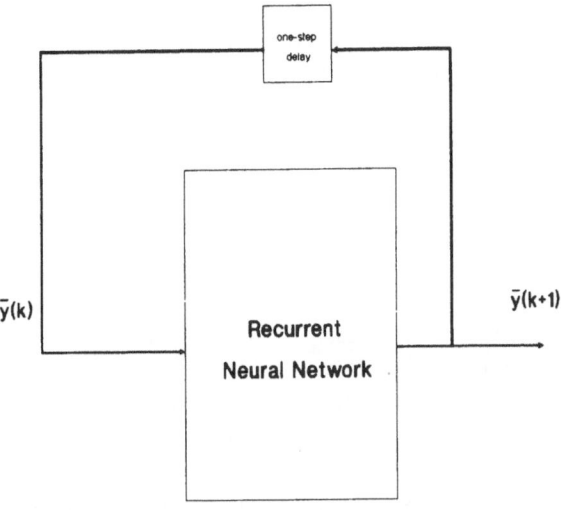

Figure 5.4 (b) Prediction using recurrent neural networks: Multi-step.

5.3 Data Sets Adopted for Simulation

Three sets of real data are employed as examples in the investigation. These data sets are shown in Tables 5.1 - 5.3. The data in Table 5.1 are the weekly averaged exchange rates between two main currencies - the British pound and the US dollar - within the period 31 December 1979 to 26 December 1983. There are 209 data in Table 5.1. They are taken from the *Exchange Rate Movement Year Book 1980-1983* [*Henley Centre for Forecasting*, 1981, 1982, 1983, 1984] which provides the weekly averaged exchange rates between all the main international currencies in the above time period. The data of Tables 5.2 and 5.3 are obtained from the Financial Times newspaper. The data of Table 5.2 are the daily exchange rates between the British pound and the Deutsche Mark from 28 October 1991 to 8 October 1992. Table 5.3 gives the daily exchange rates between the British pound and the US dollar in the same time period. Both Table 5.2 and Table 5.3 contain 240 data.

Table 5.1 Weekly averaged exchange rates between the British pound and the US dollar (Period 31/12/79 - 26/12/83) (Lesseps, M. and Morrel, J. G. , Forecasting Exchange Rates: Theory and Practice, Henley Centre for Forecasting, 1977)

1	2.23487	43	2.43670	85	1.79500	127	1.79362	169	1.46500
2	2.26150	44	2.43800	86	1.84390	128	1.77760	170	1.46412
3	2.27680	45	2.44010	87	1.84370	129	1.75060	171	1.50725
4	2.27620	46	2.40420	88	1.84100	130	1.73170	172	1.53870
5	2.26070	47	2.38210	89	1.79540	131	1.73190	173	1.55040
6	2.29740	48	2.35770	90	1.82970	132	1.71900	174	1.56500
7	2.30370	49	2.34700	91	1.81460	133	1.72510	175	1.57875
8	2.28220	50	2.33210	92	1.80610	134	1.74880	176	1.56740
9	2.27980	51	2.32680	93	1.87480	135	1.74860	177	1.55690
10	2.23650	52	2.36650	94	1.85330	136	1.73130	178	1.58270
11	2.22310	53	2.38283	95	1.82380	137	1.69960	179	1.58450
12	2.18990	54	2.41120	96	1.82650	138	1.72020	180	1.57370
13	2.18400	55	2.39570	97	1.87410	139	1.75430	181	1.52980
14	2.15012	56	2.41305	98	1.88830	140	1.72562	182	1.53530
15	2.18000	57	2.39895	99	1.90600	141	1.72140	183	1.53320
16	2.21010	58	2.34850	100	1.93000	142	1.70830	184	1.53600
17	2.25530	59	2.32325	101	1.94650	143	1.71174	185	1.52780
18	2.26554	60	2.27660	102	1.91210	144	1.69550	186	1.52220
19	2.27882	61	2.22780	103	1.88130	145	1.69465	187	1.52410
20	2.28722	62	2.19665	104	1.88425	146	1.71170	188	1.49890
21	2.31610	63	2.21970	105	1.90467	147	1.70070	189	1.48580
22	2.35525	64	2.26270	106	1.92040	148	1.67980	190	1.51110
23	2.32545	65	2.25140	107	1.87210	149	1.67530	191	1.51730
24	2.34000	66	2.23030	108	1.88330	150	1.65740	192	1.49687
25	2.33490	67	2.18850	109	1.86910	151	1.61690	193	1.49510
26	2.34110	68	2.16350	110	1.86280	152	1.59305	194	1.49720
27	2.35644	69	2.17387	111	1.84810	153	1.62970	195	1.50510
28	2.37650	70	2.14600	112	1.84210	154	1.62140	196	1.49960
29	2.37540	71	2.11250	113	1.83290	155	1.61240	197	1.48970
30	2.38826	72	2.08190	114	1.82500	156	1.60570	198	1.50425
31	2.35640	73	2.07638	115	1.81540	157	1.62200	199	1.50060
32	2.36310	74	2.06650	116	1.80600	158	1.61637	200	1.49690
33	2.37570	75	1.99670	117	1.80020	159	1.58200	201	1.48920
34	2.36690	76	1.95240	118	1.78340	160	1.57530	202	1.48630
35	2.38925	77	1.98860	119	1.75937	161	1.53900	203	1.48060
36	2.41486	78	1.97560	120	1.76150	162	1.52070	204	1.46670
37	2.40980	79	1.91060	121	1.77050	163	1.54040	205	1.45910
38	2.38940	80	1.89090	122	1.78550	164	1.54280	206	1.44280
39	2.39930	81	1.88080	123	1.81206	165	1.52660	207	1.42080
40	2.38900	82	1.85880	124	1.83120	166	1.51070	208	1.42410
41	2.39680	83	1.85588	125	1.80100	167	1.50640	209	1.44500
42	2.41220	84	1.79800	126	1.79910	168	1.50440		

Table 5.2 Daily exchange rates between the British pound and the Deutsche Mark (28/10/91 - 8/10/92), (Financial Times)

1	2.910	41	2.840	81	2.880	121	2.917	161	2.920	201	2.817
2	2.913	42	2.850	82	2.880	122	2.933	162	2.920	202	2.815
3	2.907	43	2.848	83	2.885	123	2.930	163	2.915	203	2.817
4	2.905	44	2.838	84	2.885	124	2.925	164	2.917	204	2.812
5	2.905	45	2.837	85	2.878	125	2.937	165	2.915	205	2.812
6	2.905	46	2.855	86	2.880	126	2.937	166	2.918	206	2.807
7	2.908	47	2.855	87	2.880	127	2.932	167	2.913	207	2.797
8	2.908	48	2.850	88	2.883	128	2.932	168	2.893	208	2.800
9	2.900	49	2.845	89	2.870	129	2.930	169	2.900	209	2.787
10	2.903	50	2.835	90	2.870	130	2.930	170	2.900	210	2.795
11	2.905	51	2.840	91	2.865	131	2.925	171	2.903	211	2.790
12	2.900	52	2.835	92	2.868	132	2.925	172	2.895	212	2.787
13	2.900	53	2.850	93	2.870	133	2.940	173	2.890	213	2.782
14	2.895	54	2.848	94	2.860	134	2.945	174	2.885	214	2.787
15	2.898	55	2.850	95	2.855	135	2.942	175	2.872	215	2.800
16	2.878	56	2.855	96	2.858	136	2.940	176	2.880	216	2.800
17	2.875	57	2.860	97	2.860	137	2.942	177	2.872	217	2.795
18	2.878	58	2.860	98	2.858	138	2.937	178	2.852	218	2.787
19	2.867	59	2.868	99	2.858	139	2.937	179	2.850	219	2.787
20	2.850	60	2.865	100	2.860	140	2.927	180	2.845	220	2.787
21	2.847	61	2.865	101	2.865	141	2.925	181	2.855	221	2.790
22	2.840	62	2.872	102	2.862	142	2.932	182	2.847	222	2.812
23	2.858	63	2.867	103	2.860	143	2.932	183	2.843	223	2.780
24	2.860	64	2.872	104	2.862	144	2.937	184	2.843	224	2.750
25	2.870	65	2.877	105	2.857	145	2.942	185	2.833	225	2.640
26	2.857	66	2.877	106	2.860	146	2.945	186	2.838	226	2.612
27	2.857	67	2.868	107	2.855	147	2.937	187	2.850	227	2.538
28	2.862	68	2.868	108	2.850	148	2.935	188	2.843	228	2.545
29	2.850	69	2.870	109	2.845	149	2.935	189	2.843	229	2.565
30	2.845	70	2.870	110	2.835	150	2.915	190	2.845	230	2.540
31	2.843	71	2.868	111	2.837	151	2.917	191	2.843	231	2.545
32	2.850	72	2.868	112	2.842	152	2.915	192	2.843	232	2.510
33	2.853	73	2.865	113	2.843	153	2.915	193	2.840	233	2.520
34	2.860	74	2.875	114	2.858	154	2.915	194	2.835	234	2.518
35	2.875	75	2.873	115	2.885	155	2.920	195	2.825	235	2.475
36	2.875	76	2.880	116	2.915	156	2.915	196	2.828	236	2.430
37	2.872	77	2.880	117	2.913	157	2.920	197	2.825	237	2.392
38	2.870	78	2.885	118	2.920	158	2.918	198	2.827	238	2.447
39	2.858	79	2.883	119	2.910	159	2.915	199	2.827	239	2.472
40	2.855	80	883	120	2.910	160	2.925	200	2.822	240	2.490

Table 5.3 Daily exchange rates between the British pound and the US dollar (28/10/91 - 8/10/92) (Financial Times)

1	1.704	41	1.877	81	1.748	121	1.748	161	1.866	201	1.931
2	1.723	42	1.876	82	1.757	122	1.763	162	1.860	202	1.916
3	1.743	43	1.878	83	1.746	123	1.765	163	1.861	203	1.926
4	1.743	44	1.867	84	1.755	124	1.772	164	1.863	204	1.929
5	1.749	45	1.871	85	1.757	125	1.779	165	1.878	205	1.935
6	1.781	46	1.868	86	1.756	126	1.775	166	1.892	206	1.939
7	1.770	47	1.847	87	1.740	127	1.770	167	1.892	207	1.947
8	1.772	48	1.877	88	1.724	128	1.774	168	1.907	208	1.995
9	1.780	49	1.873	89	1.717	129	1.785	169	1.904	209	1.991
10	1.768	50	1.872	90	1.719	130	1.785	170	1.911	210	1.987
11	1.769	51	1.794	91	1.723	131	1.782	171	1.918	211	1.978
12	1.773	52	1.803	92	1.716	132	1.794	172	1.909	212	1.982
13	1.771	53	1.792	93	1.725	133	1.796	173	1.910	213	1.997
14	1.775	54	1.756	94	1.709	134	1.787	174	1.926	214	2.004
15	1.769	55	1.761	95	1.713	135	1.793	175	1.919	215	1.984
16	1.791	56	1.777	96	1.716	136	1.812	176	1.896	216	1.995
17	1.796	57	1.791	97	1.732	137	1.813	177	1.921	217	1.993
18	1.796	58	1.807	98	1.733	138	1.822	178	1.929	218	2.004
19	1.797	59	1.807	99	1.712	139	1.819	179	1.913	219	1.975
20	1.798	60	1.796	100	1.707	140	1.835	180	1.926	220	1.977
21	1.799	61	1.806	101	1.715	141	1.839	181	1.937	221	1.924
22	1.797	62	1.776	102	1.720	142	1.830	182	1.949	222	1.892
23	1.767	63	1.795	103	1.731	143	1.817	183	1.918	223	1.871
24	1.768	64	1.806	104	1.722	144	1.814	184	1.909	224	1.811
25	1.765	65	1.774	105	1.738	145	1.828	185	1.904	225	1.781
26	1.776	66	1.790	106	1.731	146	1.801	186	1.914	226	1.740
27	1.771	67	1.800	107	1.735	147	1.806	187	1.900	227	1.712
28	1.773	68	1.801	108	1.724	148	1.823	188	1.916	228	1.709
29	1.793	69	1.808	109	1.730	149	1.825	189	1.929	229	1.709
30	1.810	70	1.817	110	1.743	150	1.816	190	1.917	230	1.710
31	1.807	71	1.838	111	1.750	151	1.825	191	1.919	231	1.715
32	1.808	72	1.826	112	1.749	152	1.835	192	1.920	232	1.730
33	1.810	73	1.802	113	1.749	153	1.834	193	1.924	233	1.770
34	1.809	74	1.789	114	1.767	154	1.833	194	1.922	234	1.782
35	1.817	75	1.771	115	1.764	155	1.833	195	1.914	235	1.737
36	1.822	76	1.769	116	1.762	156	1.846	196	1.908	236	1.726
37	1.820	77	1.774	117	1.772	157	1.852	197	1.930	237	1.700
38	1.825	78	1.752	118	1.756	158	1.857	198	1.927	238	1.712
39	1.834	79	1.751	119	1.748	159	1.860	199	1.921	239	1.717
40	1.860	80	1.747	120	1.748	160	1.854	200	1.928	240	1.690

5.4 Prediction Based on GMDH Networks

The GMDH network [Pham and Liu, 1994] is applied to the above data sets to carry out predictions.

In the simulation for data set 1, the exchange rates for the past three weeks are used as inputs to the network and the network is trained to predict the rate for the fourth week. For the 209 available data, the first 150 data are used for training and selection (training set: data 1, 3, 5, ... , 149; selection set: data 2, 4, 6, ... , 150). The remaining 59 data are retained to test the performance of the trained network to unseen data. The learning rate adopted in the simulation is 0.1. The first layer initially has three processing elements and they are trained for 856 epochs. All the three processing elements in the first layer are kept after training and selection. The second layer again has three processing elements and is trained for 31 epochs. All the processing elements of layer 2 also pass the selection, leading to the third layer with three initial processing elements. It is found that the RMS error given by the best processing element in the third layer is larger than that for the best processing element in the second layer, therefore the second layer becomes the output layer. After trimming, the trained network is obtained as shown in Figure 5.5.

The overall 209 data are sent to the trained network at the evaluation stage. The network is tested on two tasks. The first task is one-step prediction. Each time three consecutive exchange rates from the real database are used as inputs to the network to produce a prediction for the fourth exchange rate which is recorded and compared with the actual rate. This one-step prediction process covers the whole 209 data. The test results are shown in Figure 5.6(a).

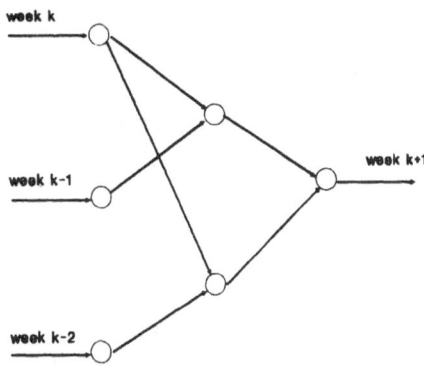

Figure 5.5 Trained GMDH network for data set 1

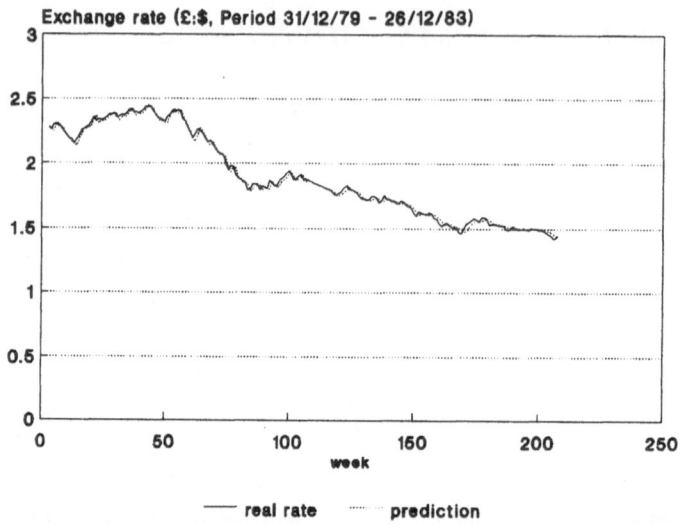

Figure 5.6 (a) GMDH prediction results for data set 1: one-step.

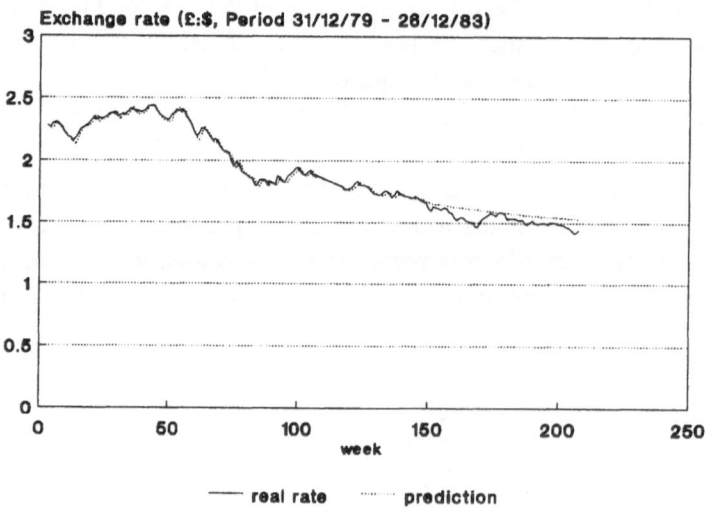

Figure 5.6 (b) GMDH prediction results for data set 1:
multi-step, starting at week 151.

The second test involves multi-step prediction. The network carries out one-step prediction until week 150. Starting from week 151, the output of the network is fedback produce a new prediction. That is, for the last 59 weeks, the network performs prediction based on its own output.   The results for the multi-step prediction are presented in Figure 5.6(b).

When simulations are carried out for data set 2, the rates for the past three days are used as inputs to the network and the network is trained to predict the rate of the fourth day.   The first 200 data of data set 2 are used for training and selection (training and selection sets are again formed by alternatively taking data from the original data set).   The other 40 data are employed as new test data.   The learning rate is 0.2.   The first layer has 3 preliminary processing elements and is trained for 3387 epochs.   Two elements are preserved to create the second layer.   The second layer has only one element and is trained for 352 epochs.   It is found that the error given by that element is smaller than that of the best element in layer 1. Therefore, the second layer becomes the output layer.   The structure of the trained network is the same as that shown in Figure 5.5 except that 'day' should be used instead of 'week'.

The same test procedure as that adopted for data set 1 is applied to data set 2.   The trained network is tested on the overall 240 data.   The results of one-step prediction are shown in Figure 5.7(a).   For multi-step prediction, the feedback of output data starts on day 201.   The results for the multi-step prediction test are given in Figure 5.7(b).

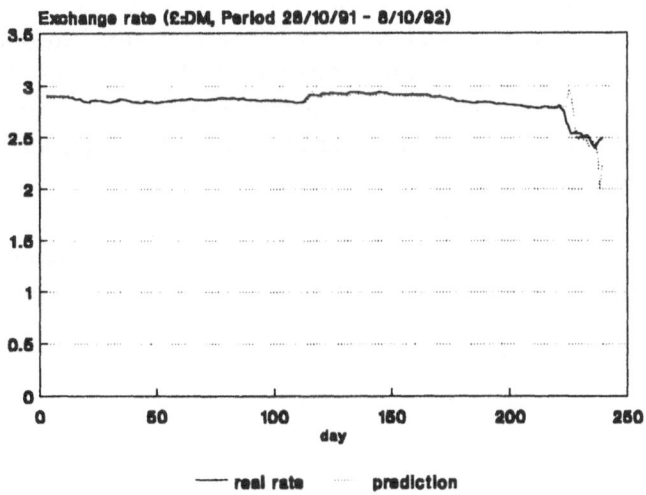

Figure 5.7 (a) GMDH prediction results for data set 2: one-step.

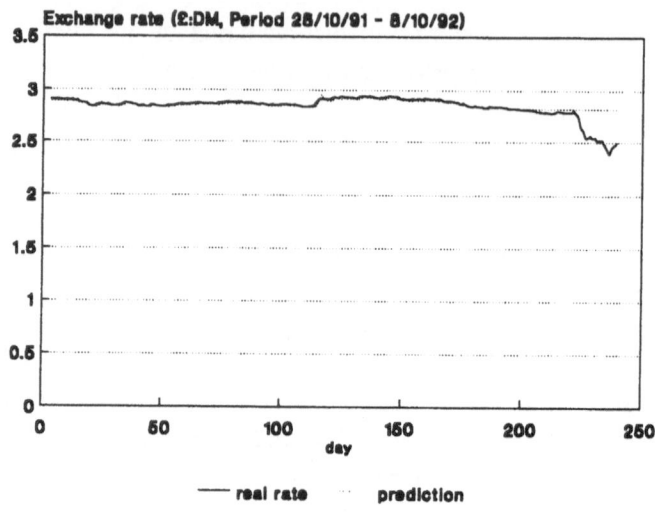

Figure 5.7 (b) GMDH prediction results for data set 2:
multi-step, starting at day 201.

In the simulation for data set 3, the rates for the past four days are used as
inputs to the network which is trained to predict the rate for the fifth day.
The learning rate is 0.04. As in the simulation for data set 2, the first 200
data are used for training and selection. These data are divided into training
and selection sets in the same manner as that used in the previous
simulations. Initially the first layer has six elements which are trained for
959 epochs. Three elements are selected from the first layer leading to three
elements in the second layer. The latter is trained for 7 epochs. All the
three elements are selected for the third layer. The minimum error in the
third layer is greater than that in the second layer, and therefore the second
layer becomes the output layer. The trained network is shown in Figure
5.8. The test results for one-step and multi-step prediction are shown in
Figures 5.9(a) and 5.9(b).

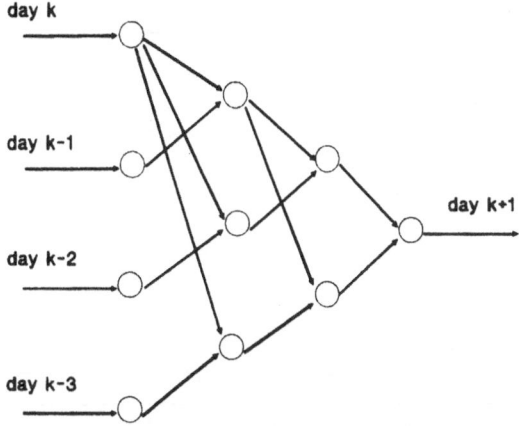

Figure 5.8 Trained GMDH network for data set 3

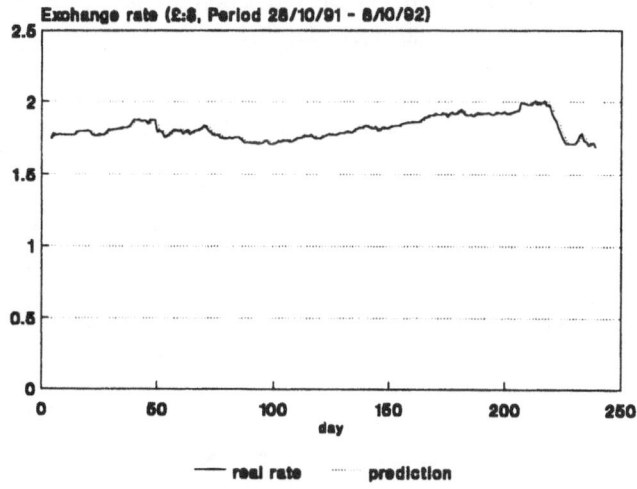

Figure 5.9 (a) GMDH prediction results for data set 3: one step.

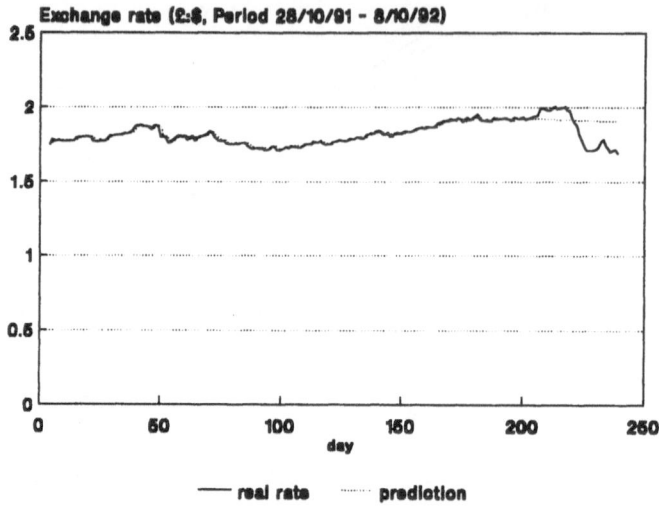

Figure 5.9 (b) GMDH prediction results for data set 3: multi-step, starting at day 201.

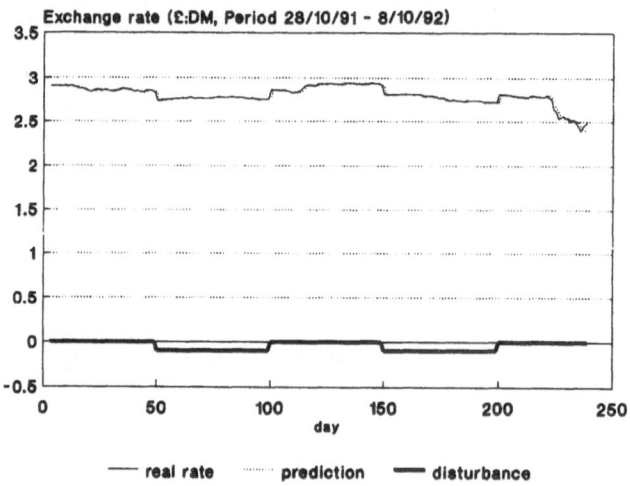

Figure 5.10 (a) GMDH prediction results for data set 2 (modified data): one-step.

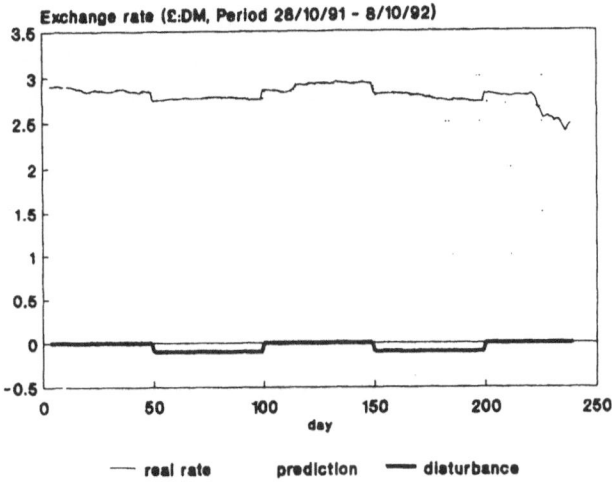

Figure 5.10 (b) GMDH prediction results for data set 2 (modified data): multi-step.

It is found from the above simulations that if there are abrupt changes in a training data set, while the major part of the data set is normal, as can be seen from Figures 5.7(a)-(b), the trained GMDH network performs satisfactorily for the normal part of the data set. However, it gives poor results for the abnormal data (data 225 to 240 in Table 5.2). To enable the trained network to accommodate data with unforeseen changes, a small disturbance in the form of step changes with magnitude 0.1 is superimposed on the data in data set 2 and the modified data set is used for training and selection. The network so trained can not only work with the normal data (data 0 to data 224), but also with the abnormal data (data 225 to data 240). The results obtained are shown in Figures 5.10(a) and 5.10(b).

5.5 Prediction Based on Multilayer Perceptron Networks

Multilayer perceptron networks are also applied to prediction tasks based on the data of Tables 5.1 to 5.3.

For data set 1, as in the GMDH network case, three input units are adopted for the MLP network. The MLP network has one hidden layer with 6 units (activation function: hyperbolic tangent) and one linear output unit.

As with the training of the GMDH network, only the first 150 data are used in the training procedure. The network is trained for 5000 epochs with the learning rate and momentum both equal to 0.1. The one-step and multi-step prediction results are shown in Figures 5.11(a)-(b).

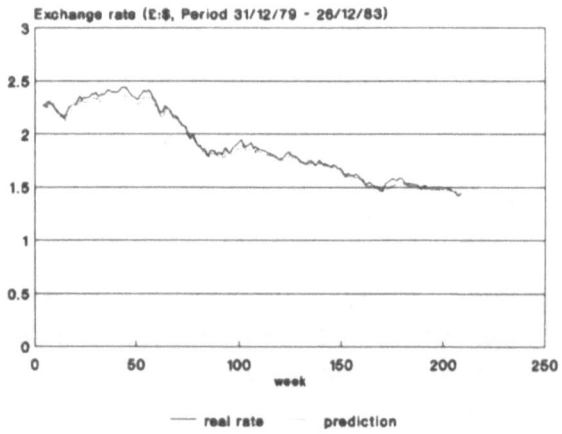

Figure 5.11 (a) MLP prediction results for data set 1: one-step.

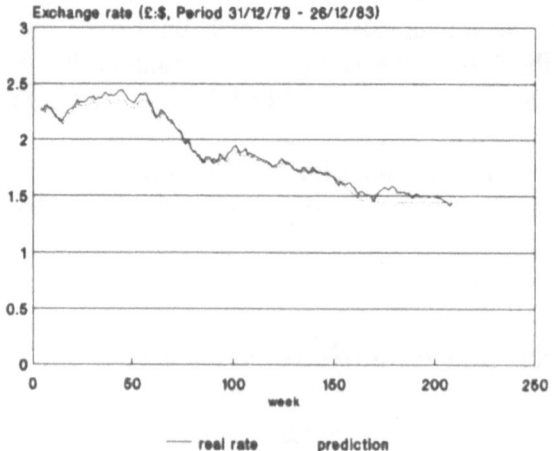

Figure 5.11 (b) MLP prediction results for data set 1: multi-step, starting at week 151.

The multilayer perceptron network for data set 2 has 3 input units, 4 hidden units and one output unit. The first 200 data are employed for training. The learning rate and momentum are both 0.1 and the network is trained for 10000 epochs. The results for one-step and multi-step prediction are shown in Figures 5.12(a)-(b).

A network of 4 input units and 4 hidden units is employed for data set 3. The network is also trained for 10000 epochs using the first 200 data with the learning rate and momentum set to 0.1 and tested on the whole 240 data. The results obtained are shown in Figures 5.13(a) and 5.13(b).

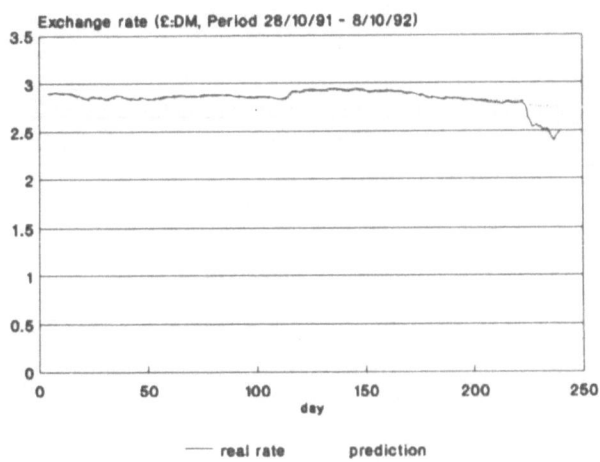

Figure 5.12 (a) MLP prediction results for data set 2: One-step.

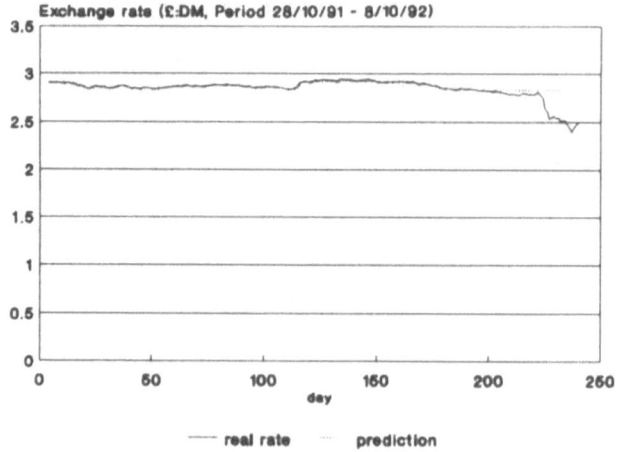

Figure 5.12 (b) MLP prediction results for data set 2:
Multi-step (starting at day 201).

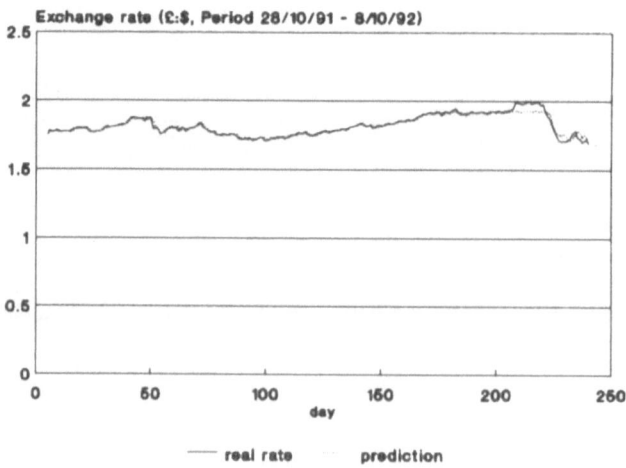

Figure 5.13 (a) MLP prediction results for data set 3: one-step.

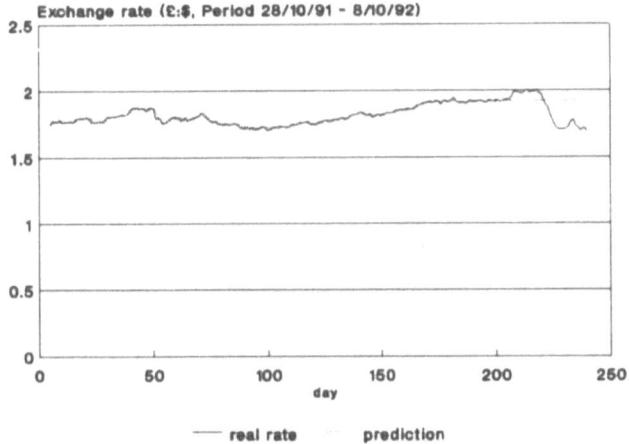

Figure 5.13 (b) MLP prediction results for data set 3: multi-step, starting at day 201.

5.6 Prediction Based on Recurrent Networks

The modified Elman network [Pham and Liu, 1992; Liu, 1993] is also employed in the above prediction examples. As pointed out in the previous sections, because the networks are recurrent, in the simulations of this section, only one input unit is employed in the input layer. Each time the prediction of the following rate is explicitly based on the present rate only. This is different from a feedforward network which needs a number of input units in the input layer to provide a prediction for the following rate based on the present and past rates.

For data set 1, the modified Elman network has 1 input unit, 5 hidden units (with hyperbolic tangent activation function), and 1 output unit. The network is trained for 20000 epochs using the first 150 data of the data set with the learning rate and momentum both equal to 0.01. The results are depicted in Figures 5.14(a)-(b).

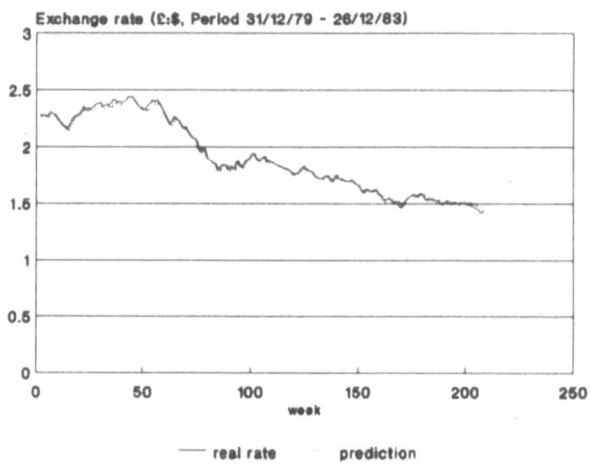

Figure 5.14 (a) Recurrent net prediction results for data set 1: one-step.

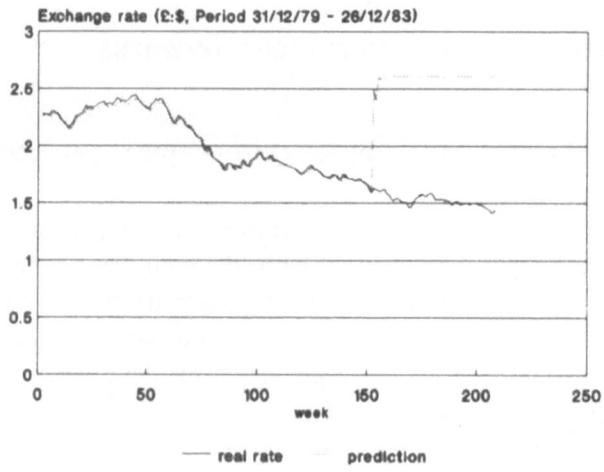

Figure 5.14 (b) Recurrent net prediction results for data set 1: multi-step, starting at week 151.

The modified Elman network for data set 2 has 1 input unit, 5 hidden units, and 1 output unit. The first 200 data from the data set are used for training. The learning rate and momentum are both equal to 0.01 and the network is trained for 1000 epochs. The results are shown in Figures 5.15(a)-(b).

The simulation for data set 3 uses a modified Elman network with 1 input unit, 5 hidden units, and 1 output unit. The network is trained for 1000 epochs with the learning rate and momentum both equal to 0.005. The first 200 data of data set 3 are employed for training. The results for one-step and multi-step predictions are presented in Figures 5.16(a)-(b).

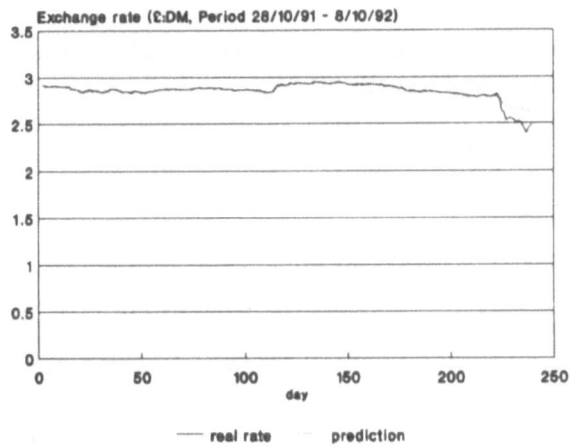

Figure 5.15 (a) Recurrent net prediction results for data 2: one-step.

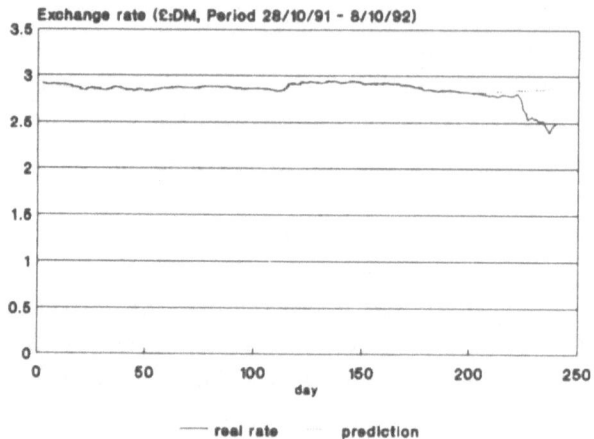

Figure 5.15 (b) Recurrent net prediction results for data 2: multi-step, starting at day 201.

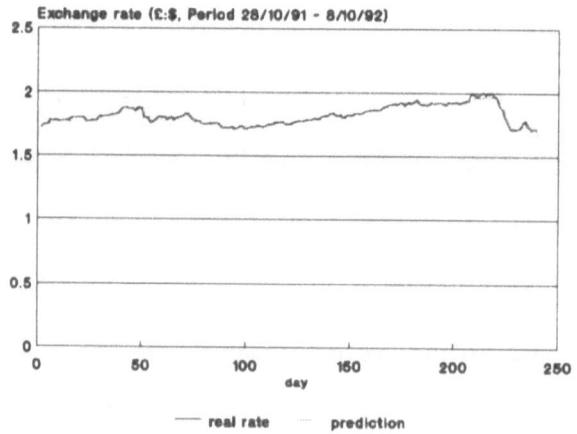

Figure 5.16 (a) Recurrent net prediction results for data set 3: One-step.

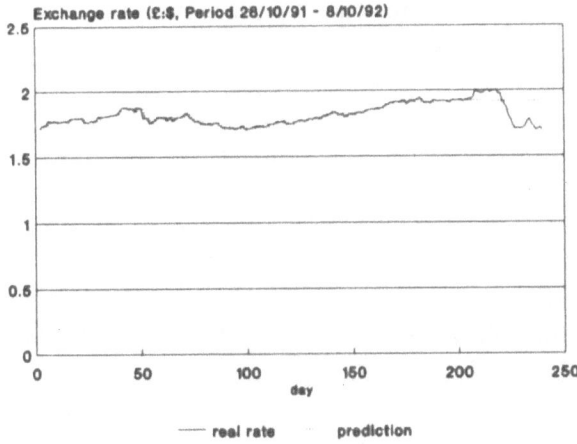

Figure 5.16 (b) Recurrent net prediction results for data set 3:
Multi-step (starting at day 201).

For comparison, the RMS errors of the GMDH network, MLP network,
and Elman network for data sets 1 - 3 are shown in Table 5.4.

Table 5.4 RMS errors of GMDH, MLP and Elman networks for data sets 1 to 3

	Data Set 1			Data Set 2			Data Set 3		
	GMDH	MLP	Elman	GMDH	MLP	Elman	GMDH	MLP	Elman
1-step	0.024	0.037	0.026	0.049	0.065	0.039	0.016	0.029	0.016
m-step	0.041	0.049	0.567	0.084	0.085	0.091	0.052	0.058	0.055

(m: 59 for data set 1, 40 for data sets 2 and 3)

5.7 Discussion

As explained in the previous sections, for the available data sets, the last 59
data (data set 1) or 40 data (data sets 2 and 3) are not employed in the

training (and selection) of the networks. To the trained networks, they are completely new data. Both the one-step prediction and multi-step prediction results for the trained networks reflect their ability to predict using new data.

From Figures 5.6(a), 5.11(a), and 5.14(a) and Table 5.4, it can be seen that for the data of Table 5.1, although all the three networks give reasonable one-step prediction results, the GMDH network has the best performance. The recurrent network does not have as good a performance as the GMDH network but is still better than the MLP network. As for the multi-step prediction tests (Figures 5.6(b), 5.11(b), and 5.14(b)), the GMDH network still gives the best results. It is able to predict the declining trend in the last 59 data which are new to it. The MLP network produces fair predictions for data 150 to 170. Nevertheless, it fails to predict the trend for the rest of the data. The recurrent network has the worst performance and cannot predict the trend at all. The output of the recurrent network saturates at a large positive value (quite a few simulations are carried out to try to improve the performance of the recurrent network, but for the data of Table 5.1, the output of the recurrent network always runs into saturation when it is employed for multi-step prediction).

Figures 5.7(a), 5.12(a), and 5.15(a) show that all the three networks are reasonably good for one-step predictions up to data 224 of data set 2. Nonetheless they are unable to provide good predictions for data 224 to 240 which suggests there is something unusual with the data. The output of the GMDH network fluctuates rapidly when there are abrupt changes in the real data. The recurrent network predicts a decreasing trend in the real data, however there are large errors in the predictions. The MLP network gives only a slight indication of the declining trend. In the multi-step prediction test (Figures 5.7(b), 5.12(b), and 5.15(b)), the GMDH network and the MLP network gives similar performances. They produces almost constant outputs for the last 40 days. Such a trend is implied in the first 200 data. Therefore, the outputs are a natural extrapolation of the historical data. The recurrent network wrongly predicts a rising trend. All of the networks fail to follow the sudden drop that started from day 225.

The GMDH network and recurrent Elman network have similarly good one-step prediction results for the data of Table 5.3, while the MLP network has the worst performance (Figures 5.9(a), 5.13(a), 5.16(a), and Table 5.4). In the multi-step prediction task, the GMDH network is the best because it predicts the declining trend of the last part of the data set, even though the predicted decline is slight compared to the real data (Figure 5.9(b)). The

MLP and recurrent networks are unable to predict the trend (Figures 5.13(b), and 5.16(b)).

Figures 5.10(a) and 5.10(b) indicate that when small step disturbances are superimposed onto the training data, the trained GMDH network is able to accommodate abrupt changes in the real data. For the abnormal data, the predictor can make good one-step predictions. In the multi-step case, the network successfully predicts the declining trend in the last 40 data.

The above observations suggest that all the neural networks tested are useful for prediction tasks. The GMDH network is the best network as it gives reasonably good predictions in most circumstances. According to [Refenes, et al, 1993], MLP networks are better than conventional prediction techniques such as ARIMA and exponential smoothing. Thus it can be expected that the GMDH network will be better than these conventional methods. The simulation results also indicate that the modification of training data sets makes the predictors capable of dealing with abnormal variations in real exchange rate systems. This suggests that modifying the training data can help to produce a more robust predictor.

5.8 Summary

This chapter has described an application of feedforward neural networks to stock market prediction. The results show that such neural networks can successfully be employed for this task. Using modular neural networks increases the prediction performance.

The chapter has also presented the results obtained with three types of neural networks in predicting exchange rates. The networks are all able to carry out reasonable one-step predictions when the economy is under normal conditions. For multi-step prediction tasks, the GMDH network provides better results than the other two networks in most circumstances. Modifying the training data enables the networks to operate with abnormal data and leads to more robust predictors.

References

Brown, R.G. (1963) *Smoothing, Forecasting and Prediction of Discrete Time Series*, New Jersey: Prentice Hall.

Hanke, J.E. and Reitsch, A.G. (1989), *Business Forecasting*, London: Allyn and Bacon.

Exchange Rate Movement Year Book 1980-1983, Henley Centre for Forecasting, London.

Kamijo, K. and Tanigawa, T. (1990) Stock price pattern recognition - A recurrent neural network approach, *1990 Int. J. Conf. on Neural Networks*, San Diego, **1**, 215-221.

Kimoto, T. and Asakawa, K. (1990) Stock market prediction system with modular neural networks, *1990 Int. J. Conf. on Neural Networks*, San Diego, **1**, 1-6.

Lee, C.H. and Park, K.C. (1992), Prediction of monthly transition of the composition stock price index using recurrent back-propagation, *Int. Conf. on Artificial Neural Networks*, 1992, Brighton, UK, 1629-1632.

Lesseps, M. and Morrel, J.G. (1977) *Forecasting Exchange Rates: Theory and Practice*, Henley Centre for Forecasting, London.

Liu, X. (1993) Dynamic system identification and prediction using neural networks, PhD thesis, School of Engineering, University of Wales, UK.

Pham, D.T. and Liu, X. (1995) Modelliing and prediction using GMDH neural networks, *Int. J. Systems Science*, in press.

Pham, D.T. and Liu, X. (1992) Dynamic system modelling using partially recurrent neural networks, *J. of Systems Engineering*, **2**(2), 90-97.

Refenes, A.N., Azema-Barac, M., Chen, L., and Karoussos, S.A., (1993) Currency exchange rate prediction and neural network design strategies, *Neural Computing and Applications*, **1**(1), 46-58.

White, H. (1988) Economic prediction using neural networks: The case of IBM daily stock returns, *1988 IEEE Int. Conf. on Neural Networks*, San Diego, CA, **2**, 451-458.

Chapter 6 Neural Network Controllers

Neural networks are developed by morphologically and computationally simulating a human brain. Although, as seen in previous chapters, the precise operation details of artificial neural networks are quite different from human brains, they are similar in three aspects. First, a neural network consists of a very large number of simple processing elements (the neurons). Second, each neuron is connected to a large number of other neurons. Third, the functionality of the networks is determined by modifying the strengths of connections during a learning phase [Psaltis et al, 1988; Hunt et al, 1992; Warwick et al, 1992]. Efforts have been made to find efficient approaches for control from physiological studies of the brain. Research over the last twenty years has revealed the architecture and performance characteristics of the brain as a controller [Albus, 1975; Ito, 1984; Kawato et al, 1987] and has shown that neural network controllers have important advantages over conventional controllers. The first advantage is that a neural network controller can efficiently utilise a much larger amount of sensory information in planning and executing a control action than an industrial controller can. The second advantage is that a neural network controller has the collective processing capability that enables it to respond quickly to complex sensory inputs while the execution speed of sophisticated control algorithms in a conventional controller is severely limited. The last but also the most important advantage of a neural network controller is that good control can be achieved through learning [Psaltis, et al, 1988]. Three controllers have played important roles in research on neural control. They are: Albus's cerebellar model articulation controller (CMAC) [Albus, 1975], Kawato et al's hierarchical neural network controller [Kawato et al, 1987], and Psaltis et al's multilayered neural network controller [Psaltis et al, 1988]. Both CMAC and the hierarchical neural network controller were developed based on physiological research on the brain, but they are quite different in architecture and methodology. The multilayered neural network approach of Psaltis et al offers important architectures for control. Each of the three controllers has its own architectural features and properties. These are analysed and compared in this chapter which also gives a comparison of neural network controllers and conventional adaptive controllers.

The neural controllers described by Kawato et al, Albus and Psaltis et al could all be regarded as inverse controllers as they are all based on modelling the inverse dynamics of the plant. Another important neural net based inverse controller, not covered in this chapter, is the internal model controller [Hunt and Sbarbaro, 1991]. In addition to these inverse-model-based controllers, there are also other neural controllers originating from conventional control approaches, for example, those based upon the variable structure technique, the robust control strategy, the model predictive method and the model reference PID-like controller. For information on these important controllers, see [Colina-Morles and Mort, 1993], [Rovithakis and Christodoulou, 1994], [Saint-Donat et al, 1991], [Evans et al, 1993] and [Lee et al, 1994].

6.1 Neural Network Controllers

6.1.1 CMAC

CMAC was developed by Albus [Albus, 1975, 1979]. It is one of the main neural controllers found in recent applications [Miller, 1989, 1990; Handelman et al, 1989; Kraft and Campagna, 1989; An et al, 1994; Wang et al, 1994; Xu et al, 1994]. A CMAC module is shown in Figure 6.1.

In Figure 6.1, S is the input state space, the sub-vectors s_i in it are input vectors of dimension n. A is an N dimensional memory. Each s in S is mapped to C locations in A. In general, the theoretical size (number of memory locations) of A is unpractically large, while the memory requirement for a typical control problem is much smaller. For this reason, the large memory A is randomly mapped into a smaller memory A', but there are still C locations in A' which correspond to each point in S. The values stored at these locations are summed to produce the CMAC output $f(S)$. As a whole, CMAC acts as an arbitrary function f such that,

$$u = f(S) \tag{6.1}$$

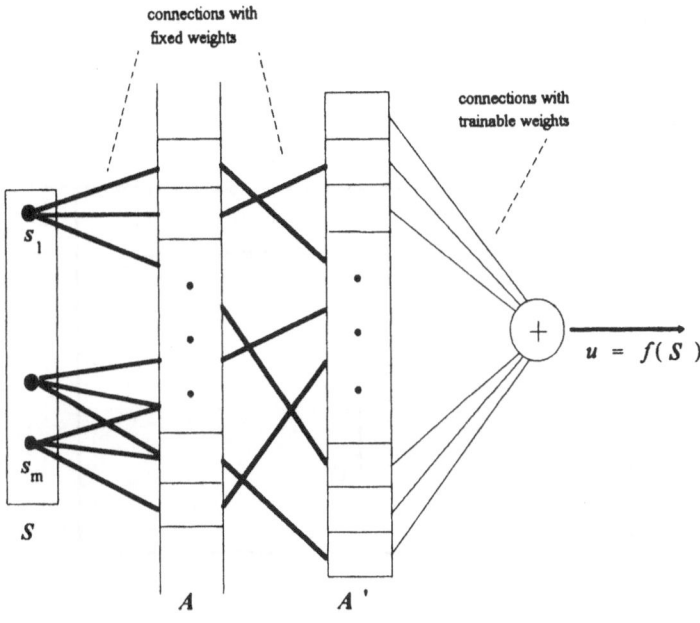

Figure 6.1 A CMAC module

where S is the input to CMAC and u is the output of CMAC. A CMAC module has a number of characteristics. Because of its input mapping, it has local input generalisation, that is, similar inputs will produce similar outputs. A large CMAC network can be used and trained in practical time. Due to the learning rule used in CMAC, it has a unique minimum. Finally, CMAC can learn a wide variety of functions. The training of CMAC can be conducted as follows: (i) Assume that f is a function for CMAC to learn. Then $u = f(.)$ is the desired CMAC output; (ii) Select a point s in the input space where u is to be stored. Compute the current value $\underline{u} = f(s)$; (iii) If $|u-\underline{u}| < \xi$, where ξ is an acceptable error, then do nothing. If $|u-\underline{u}| > \xi$, then add to every connection weight which contributed to u the quantity $\Delta = \alpha$ $(u-\underline{u})/|A^*|$, where $|A^*|$ = the number of weights from A' which contributed to u and α is the learning rate, a quantity between zero and one; (iv) Select another input point, repeat the above procedure until all the input points have been processed.

There are two schemes for using CMAC for control. The first scheme has the configuration depicted in Figure 6.2 [Albus, 1975].

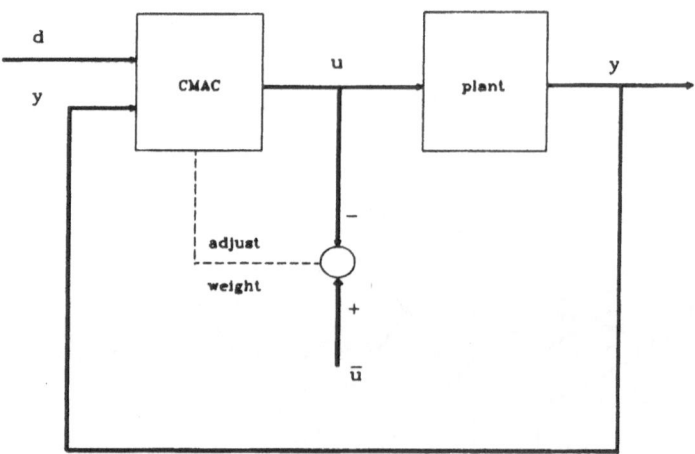

Figure 6.2 CMAC based control system (scheme 1)

Figure 6.2 shows that, in the control system, both the command and feedback signals are used as inputs to the CMAC controller. The output of the controller is fed directly to the plant. The desired output of the neural network controller has to be supplied. The training of the controller is based on the error between the desired and actual controller output. Two stages are needed to make the system work. The first stage is training the controller. When CMAC receives the command and feedback signals, it produces an output. This output is compared with the desired output. If there are differences between them, then the weights are adjusted to eliminate the differences. On completion of this stage, CMAC has learnt how to produce a suitable output to control the plant according to the given command and the measured feedback signals. The second stage is control. CMAC can work well when the required control is close to that with which it has been trained. Both stages are completed without the need to analyse the dynamics of the plant and to solve complex equations. However, in the training stage, this scheme requires the desired plant input to be known.

The second control scheme is illustrated in Figure 6.3 [Miller, 1987, 1989].

In this scheme, the reference output block produces a desired output at each control cycle. The desired output is sent to the CMAC module which provides a signal to supplement the control signal from a fixed gain conventional error feedback controller. At the end of each control cycle, a training step is executed. The observed plant output during the previous control cycle is used as input to the CMAC module. The difference between the computed plant input u^* and the actual input u is used to compute the weight adjustment. As CMAC is trained continually following successive control cycles, the CMAC function forms an approximation of the plant inverse transfer function over particular regions of the input space. If the future desired outputs are in regions similar to previous observed outputs, the CMAC output will be similar to the actual plant input required. As a result, the output errors will be small and CMAC will take over from the fixed-gain conventional controller.

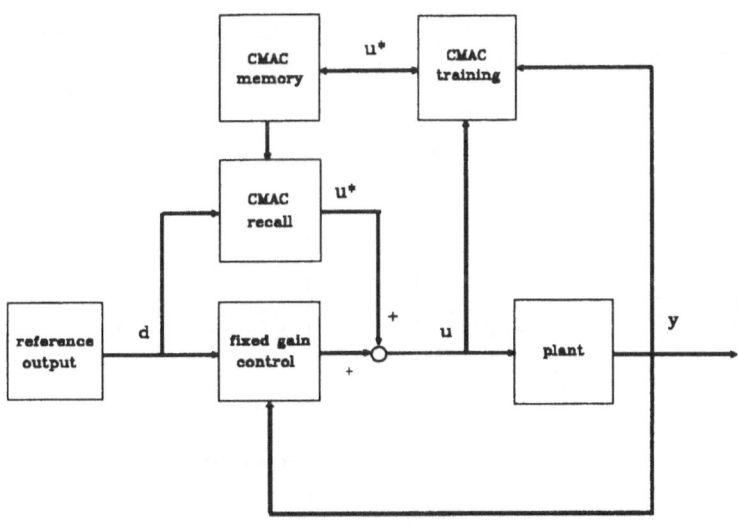

Figure 6.3   CMAC based control system (scheme 2)

From the above description, Scheme 1 is a closed-loop control system, because besides the command variables, the feedback variables are used as inputs to the CMAC module to be encoded so that any variations in the plant output can cause variations in the input it receives. In scheme 1, the adjustment of weights is based on the error between the desired controller output and the actual controller output, rather than the error between desired plant output and actual plant output. As already mentioned, this requires the designer to assign the desired controller output and will cause problems because usually only the desired plant output is known to the designer. The training in scheme 1 can be considered to be the identification of a proper feedback controller. In scheme 2, the CMAC module is used for learning an inverse transfer function with the assistance of a conventional fixed-gain feedback controller. After training, CMAC will be the principal controller. In this scheme, control and learning proceed at the same time. The disadvantage of this scheme is that it requires a fixed-gain controller to be designed for the plant.

6.1.2 Hierarchical Neural Network Model

Similar to CMAC, the hierarchical neural network model was proposed by Kawato et al based on research results from the area of physiology [Kawato, 1987, 1988]. It is depicted in Figure 6.4 (as compared with [Kawato, 1987], some changes have been made for discussion convenience).

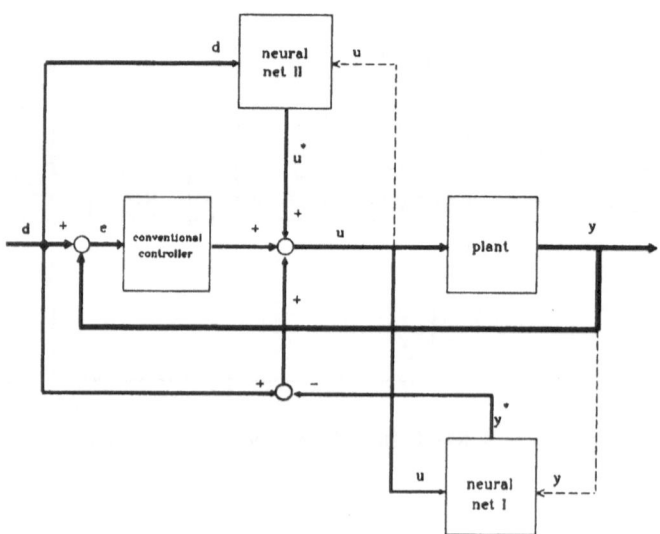

Figure 6.4 Hierarchical neural network model

In Figure 6.4, d is the desired plant output, u is the control input of the plant, y is the actual plant output. u^* and y^* are the computed plant input and output as given by the neural networks. The system can be considered to consist of three parts. The first part is the loop composed of the dark lines. This is a conventional feedback loop known as external feedback [Kawato, 1987]. The feedback control is based on the error e between the desired plant output d and the actual plant output y measured by sensors, i.e. e = (d - y). Usually the external conventional feedback controller is a proportional-derivative controller. The second part is the path connected with neural network I which is an internal model of the plant dynamics. This neural network monitors the plant input u and output y and learns the plant dynamics. After learning, it can provide an approximate plant output y^* when it receives the plant input u. In this sense, this part acts as a system dynamics identifier. Based on the error $d - y^*$, this part provides an internal feedback loop which is much faster than the external feedback loop as the latter usually has sensory delays in the feedback path [Kawato, 1987]. The third part of the system is neural network II which monitors the desired output d and the plant input u. This neural network learns to model the plant inverse dynamics. After learning, when it receives the desired output command *d*, it can produce the appropriate plant input u^*. The hierarchical neural network model controlled system operates according to the following procedure. The sensory feedback is effective mainly in the learning stage. This loop provides a conventional feedback signal to control the plant. Because of the sensory delay and thus small allowable control gain, the system response is slow, which limits the speed of the learning stage. During the learning stage, neural network I learns the system dynamics, while neural network II learns the inverse dynamics. As learning proceeds, the internal feedback gradually takes over the role of the external feedback as the main controller. Then, as learning proceeds further, the inverse dynamics part will replace the internal feedback control. The final result is that the plant is controlled mainly by a feedforward controller since the plant output error is nearly absent with the internal feedback providing fast control to deal with random disturbances. In the above procedure, control and learning are executed simultaneously. The neural networks function as identifiers: one for the identification of plant dynamics, and the other for the identification of inverse dynamics.

Importantly, a hierarchical neural network model based control system can be separated into two subsystems, the (forward) dynamics identifier based system and the inverse dynamics identifier based system, which can be applied individually. The dynamics identifier based system is depicted in

Figure 6.5 and the inverse dynamics identifier based system is illustrated in Figure 6.6.

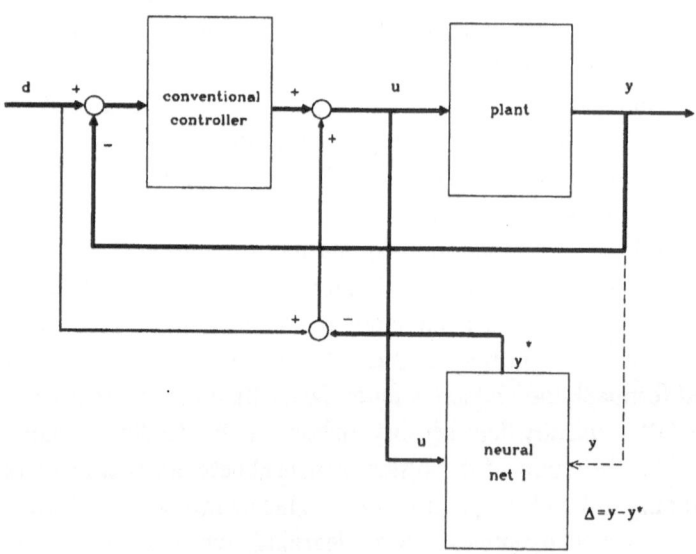

Figure 6.5 Dynamics identifier based system

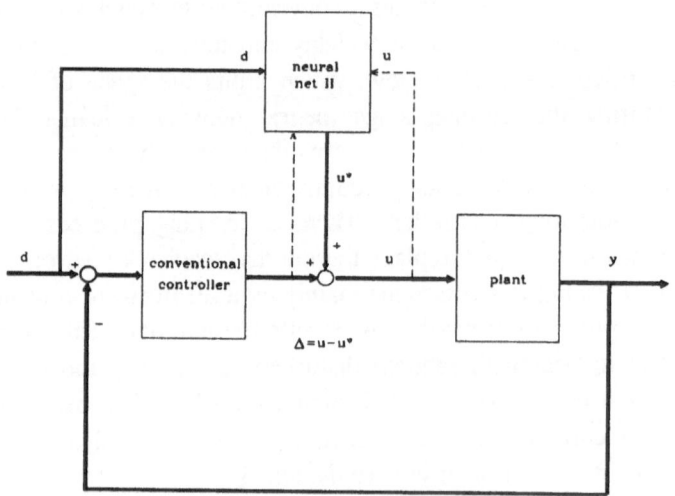

Figure 6.6 Inverse-dynamics-identifier-based system

In summary, a hierarchical neural network model based system has the following characteristics:

(i) The system has two identifiers, one for the identification of plant dynamics, the other for inverse plant dynamics;

(ii) There is a main feedback loop, which is important in the training of the neural networks;

(iii) As training proceeds, the inverse dynamics part becomes the main controller;

(iv) The final effects of hierarchical neural network model based control are similar to feedforward control.

6.1.3 Multilayered Neural Network Controller

The multilayered neural network controller proposed in [Psaltis, 1988] is essentially a feedforward controller.

Consider a general control system shown in Figure 6.7. In this system, there are two kinds of control action: feedforward and conventional feedback control. The feedforward control part is implemented by a neural network. The aim of training the feedforward part is to minimise the error between the desired and actual plant output. This error is the input to the feedback controller. Psaltis et al [Psaltis et al, 1988] considered the feedback and feedforward actions separately, concentrating on the training of the feedforward controller by ignoring the existence of the feedback control. Three learning architectures were proposed: indirect, general, and specialised.

Indirect learning architecture: The indirect learning architecture shown in Figure 6.8 has two identical neural networks for training. In this architecture, each network acts as an inverse dynamics identifier. The goal of training is to find an appropriate plant control u from the desired response d. The weights are adjusted based on the difference between the outputs of networks I and II to minimise error e, because if network I can be trained so that $y = d$, then $u = u^*$. However, this cannot guarantee that the error between the desired output d and the actual output y is minimised [Psaltis et al, 1988].

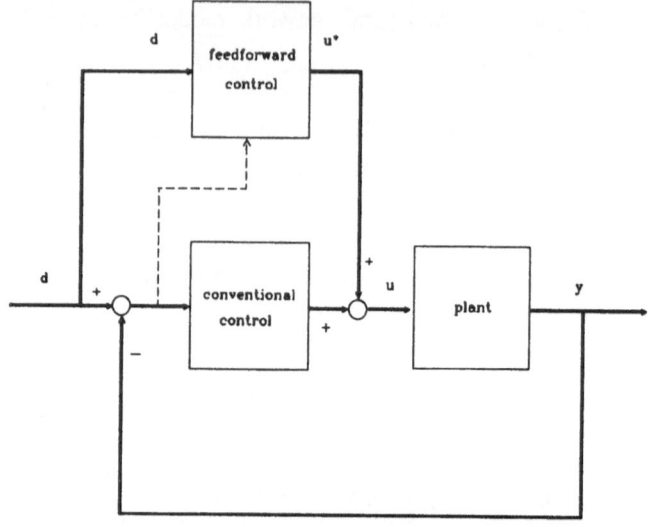

Figure 6.7 A general multilayer neural network control system

General learning architecture: The general learning architecture shown in Figure 6.9 does minimise e = d - y.

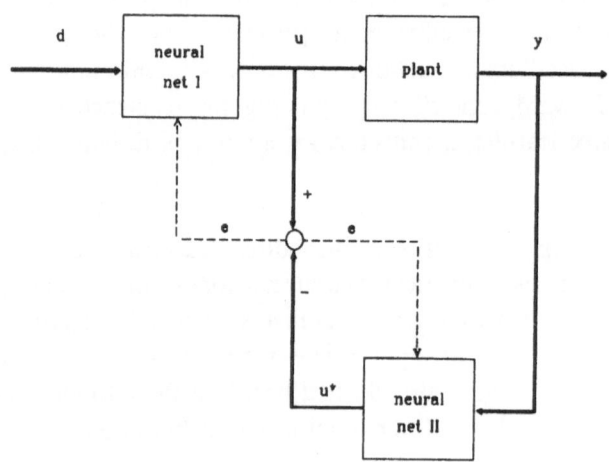

Figure 6.8 Indirect learning architecture

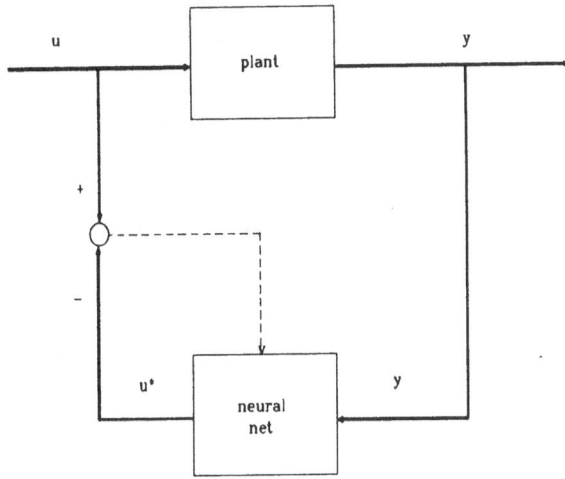

Figure 6.9 General learning architecture

In this architecture, the network is trained to minimise the error between the plant input u and the network output u*. In training, u should be in such a range that y covers the desired output d. After training, the network is able to provide an appropriate u to the plant if a desired output d is sent to it. The limitation of this architecture is that generally which u corresponds to the desired output d is not known so that the network has to be trained over a large range of u to enable the plant output y to include the desired value d during learning.

Specialised learning architecture: Figure 6.10 shows the specialised learning architecture. In this architecture, the desired output d is the input to the network when it is being trained. By applying the error back propagation method, the difference e between the desired output d and the actual output y of the plant is minimised through training. Thus, not only can a good plant output be expected but also the training can be executed in the region of the desired output, without having to know the proper range of plant input. However, in this architecture the plant is treated as a layer of the network. To be able to train the network, either the plant dynamics model has to be known or some approximation has to be made. Learning in the multilayered neural network controller is accomplished by the error back propagation training algorithm [Miller et al, 1990]. The error can be the difference between the desired and actual plant output, or the difference between the correct plant input and the input calculated by the neural networks.

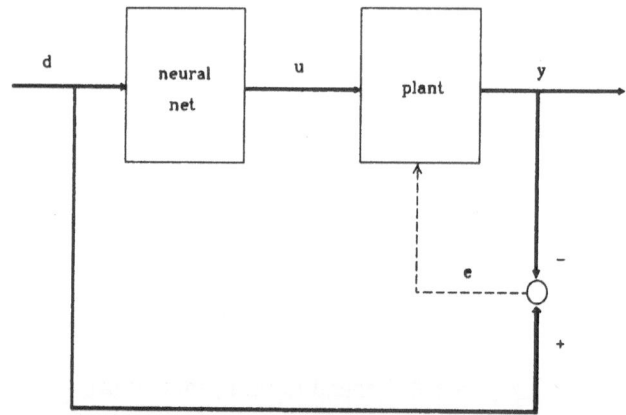

Figure 6.10 Specialised learning architecture

6.2 Comparison of Neural Network Controllers

6.2.1 Hierarchical Neural Network Model and Multilayered Neural Network Controller

By comparing the hierarchical neural network model and the multilayered neural network controller, the following can be seen:

(i) Both the hierarchical neural network model and the multilayered neural network controller use the desired plant output as the system input. Hence the whole system can be easily trained in the region of the desired output;

(ii) The main controller in both the multilayered neural network scheme and the hierarchical neural network model is a feedforward controller;

(iii)   In the multilayered neural network controller, the feedforward controller is trained without considering the feedback controller, while in the hierarchical neural network model the feedforward controller is formed with the assistance of feedback control.

6.2.2 Multilayered Neural Network Controller (or Hierarchical Neural Network Model) and CMAC

The multilayered neural network controller (or the hierarchical neural network model) and CMAC have the following contrasting features:

(i) CMAC is a table looking up method, that is, in CMAC, all the information needed to produce a proper control is stored in a table and memory size is a problem. In the multilayered neural network controller, the needed information is stored in the connections of a network. Usually, the memory size of a multilayered neural network controller for a given task is smaller than that of CMAC;

(ii) In CMAC scheme 1, the feedback signal affects the selection of the weights. Thus, the feedback signal is effective in both training and control. In the multilayered neural network controller and the hierarchical neural network model, the feedback signal is mainly involved in the training stage. After training, the feedback becomes less important and has no effects on the feedforward controller (i.e. the neural network controller) if on-line training is not applied. In this sense, CMAC scheme 1 is still a feedback controller, because even after training, the controller output still relies on sensory feedback signals, while CMAC scheme 2 is similar to the hierarchical neural network model and the multilayered neural network controller;

(iii) In the multilayered neural network controller, the neural network controller is trained based on the error between the desired plant output and the actual plant output. This presents no problem because the desired output is usually known. However, in CMAC scheme 1, the weight table is adjusted using the error between the desired plant input and the actual plant input. This sometimes causes difficulties because which plant input pattern can drive the plant output to follow a desired output is usually unknown. CMAC scheme 2 has overcome this problem;

(iv) In the hierarchical neural network model and the multilayered neural network controller, the dynamics and/or inverse dynamics of the plant are learned so that the controller can adapt to meet different control requirements. In CMAC scheme 1, a special control pattern is learned during training.   For different control requirements, retraining of the controller is needed. CMAC scheme 2 learns the inverse dynamics of the system and therefore it does not have such a problem;

(v) If on-line training is used, CMAC is faster than the hierarchical neural network model and the multilayered neural network controller because usually CMAC needs fewer iterations to train [Miller et al, 1990].

6.2.3 Common Aspects of Neural Network Controllers

The neural network controllers have the following characteristics in common:

(i)   All the neural network controllers can be realised without requiring explicit control algorithms.   The controllers achieve correct performance through learning. The controlled plants can be highly nonlinear. For plants with constant dynamics, no matter what nonlinear features they have [Li and Slotine, 1989], the neural network controllers can be trained off-line. For time-varying dynamic plants, on-line training is necessary;

(ii)   All the neural network controllers can process a large amount of information.   Increasing the input information does not significantly decrease the processing speed of the neural network controllers.   The increase affects mainly the speed of training but not so much the speed of control;

(iii) All the neural network controllers have a good ability to accommodate input noise.   In CMAC, this is because of its input encoding.   In the hierarchical neural network model and the multilayered neural network controllers, the capability comes from the filtering characteristics of neural networks.

6.2.4 Comparison with Adaptive Controllers

A neural network controller can be regarded as an adaptive controller if the weights are updated on-line.   However, there are differences between a neural network controller and conventional adaptive controllers such as the model reference adaptive system (MRAS) and self tuning regulator (STR).

CMAC and Adaptive Controllers: Architecturally, CMAC scheme 2 and adaptive controllers are similar. They are both configured to regulate the output of the feedback controller in a closed loop. The feedback controller in CMAC and adaptive control systems receive external sensory feedback signals. Such signals are used for the adaptation of controller parameters in STR and MRAS and for both training and recall in CMAC. The amount of calculation for STR and MRAS is the same during the whole control process but in CMAC it is much less in recall than in training because in recall no multiplication is needed. Therefore, after training CMAC has a better a response speed than STR and MRAS. On the other hand, simulation studies by Kraft and Campagna [Kraft and Campagna, 1989] have shown that although STR and MRAS are comparable with CMAC for noise free linear systems, a CMAC-based system is more robust to noise. CMAC is also better than STR and MRAS in the control of nonlinear systems.

Hierarchical Neural Network Model / Multilayer Neural Network Controller and Adaptive Controllers: Both MRAS and STR are designed as feedback controllers. On the other hand, as already mentioned, in the hierarchical neural network model and the multilayered neural network controller, the main controllers are feedforward. The feedback controllers are mainly for training and for the suppression of output noise. Thus, a better response speed can be expected in neural network control. Again, as mentioned previously, because of the parallel processing architecture of neural networks, the computational complexity of neural network controllers does not increase much with the complexity of the plant. Therefore, a large amount of information can be processed in a neural network control system. On the contrary, conventional adaptive controllers become very complicated when a plant has complex dynamics. The learning ability of a neural controller is more powerful than that of a conventional adaptive controller [Bavarian, 1988]. The learning ability of an adaptive controller is limited by a priori knowledge about plant structure, input signal, regulator structure etc., while a neural network controller can learn with little a priori knowledge. Moreover, a neural network controller has a generalisation capability, that is, it can operate with different input signal ranges. For an adaptive controller, reprogramming is usually necessary. Neural network control systems have greater robustness than adaptive controllers to input signals. The hierarchical neural network model also has a certain amount of robustness to variations in the parameters of plant dynamics [Kawato et al, 1987]. If a neural network controller is trained off-line, it is likely for them to have poor robustness to plant parameter variations and output noise because it has a feedforward controller as the main control. The feedback control part in both the hierarchical neural network model and the multilayered neural network controller is also helpful for the control of output noise and plant parameter variations. Moreover, the stability of adaptive control systems is dependent on the plant dynamics. In neural network control systems, because their dominant control characteristic is feedforward, stability depends mainly on the structure of the network itself [Guez et al, 1988]. The identification of parameters in adaptive control systems takes place during the whole control process. In neural network control, dynamics identification occurs only in the learning stage (except when on-line learning is implemented) because of the memorisation ability of a neural controller. Where on-line learning is employed, then identification happens continuously in a neural controller as in a conventional adaptive controller.

6.3 Summary

This chapter has reviewed three main types of neural network controllers: Albus's CMAC, Kawato's hierarchical controller and Psaltis's multilayered controller. The chapter has highlighted the main characteristics of each type and provided a comparison between the neural controllers and conventional adaptive controllers.

References

Albus, J. S. (1975) A new approach to manipulator control: cerebellar model articulation control (CMAC), *Trans. ASME, J. of Dynamics Syst., Meas. and Contr.*, 97, 220-227.

Albus, J. S. (1975) Data storage in the cerebellar model articulation controller (CMAC), *Trans. ASME, J. of Dynamics Syst., Meas. and Contr.*, 97, 228-233.

Albus, J. S. (1979) A model of the brain for robot control, *Byte*, 54-95.

Albus, J. S. (1979) Mechanisms of planning and problem solving in the brain, *Math. Biosci.*, **45**, 247-293.

An, P.E., Brown, M., Harris, C.J., Lawrence, A.J., Moore, C.J. (1994) Associative memory neural networks: adaptive modelling theory, software implementations and graphical user, *Engng. Appli. Artif. Intell.*, 7(1), 1-21.

Astrom, K.J. and Wittenmark, B. (1989) *Adaptive Control*, Reading: Addison Wesley.

Bavarian, B. (1988) Introduction to neural networks for intelligent control, *IEEE Contr. Syst. Mag.*, 3-7.

Colina-Morles, E. and Mort, N. (1993) Neural network-based adaptive control design, *Journal of Systems Engineering*, 2(1), 9-14.

Chen, F.C. (1990) Back propagation neural networks for nonlinear self tuning adaptive control, *IEEE Contr. Syst. Mag.*, 44-48.

Evans, J.T., Gomm, J.B., Williams, D., Lisboa, P.J.G. and To, Q.S. (1993) A practical application of neural modelling and predictive control, in Page, G.F., Gomm, J.B. and Williams, D., *Applications of Neural Networks to Modelling and Control*, London: Chapman & Hall, 1993.

Guez, A. Eilbert, J.L. and Kam, M. (1988) Neural network architecture for control, *IEEE Contr. Syst. Mag.*, 22-25.

Handelman, D. A., Lane, S. H. and Gelfand, J. J. (1989) Integrating neural networks and knowledge based systems for robotic control, *Proc. 1989 IEEE Int. Conf. on Robotics and Auto.*, **3**, 1454-460.

Hunt, K.J. and Sbarbaro, D. (1991) Neural networks for nonlinear internal model control, *IEE Proceedings-D*, **138**(5), 431-438.

Hunt, K.J., Sbarbaro, D., Zbikowski, R. and Gawthrop, P.T. (1992) Neural networks for control systems - a survey, *Automatica*, **28**(6), 1083-1112.

Ito, M. (1984) *The cerebellar and neural control*, New York: Raven Press.

Kawato, M., Furukawa, K. and Suzuki, R. (1987) A hierarchical neural network model for control and learning of voluntary movement, *Biol. Cybern.*, **57**, 169-185.

Kawato, M., Uno, Y., Isobe, M. and Suzuki, R. (1988) Hierarchical neural network model for voluntary movements with application to robotics, *IEEE Contr. Mag.*, 8-15.

Kraft, L. K. and Campagna, D. P. (1989) A comparison of CMAC neural network and traditional adaptive control systems, *1989 American Control Conference*, Pittsburgh, PA, June 21-23, 1, 884-889.

Kung, S.Y. and Hwang, J.N. (1989) Neural network architectures for robotic applications, *IEEE Tras. Robotics and Auto.*, 5(5),641-657.

Lee, T.H., Yue, P.K. and Tan, K.K. (1994) A neural-network-based model reference PID-like controller for process control, *Eng. Applications of Artificial Intel.*, 7(6), 677-684.

Li, W. and Slotine, J.J.E. (1989) Neural network control of unknown nonlinear systems, *1989 American Control Conference*, Pittsburgh, PA, June 21-23, 1136-1141.

Miller III, W.T., Glanz, F.H. and Kraft III, L.G. (1987) Application of a general learning algorithm to the control of robotic manipulators, *Int. J. of Robotics Research*, 6(2), 84-98.

Miller III, W. T. (1989) Real time application of neural networks for sensor based control of robots with vision, *IEEE Trans. Syst. Man, and Cybern.*, 19(4), 825-831.

Miller III, W.T., Henes, R.P., Glanz, F.H. and Kraft III, L.G. (1990) Real time dynamic control of an industrial manipulator using a neural network based learning controller, *IEEE Trans. on Robotics and Automation*, 6(1), 1-9.

Miller III, W.T., Glanz, F.H. and Kraft III, L.G. (1990) CMAC: An associative neural network alternative to backpropagation, *Proc. of IEEE*, 78(10), 1561-1567.

Miller III, W.T., Sutton, R.S. and Werbos, P.J. (eds) (1990) *Neural Networks for Control*, Cambridge, MA: MIT Press.

Miyamoto, H., Kawato, M. Setoyama, T. and Suzuki, R. (1988) Feedback error learning neural network for trajectory control of a robotic manipulator, *Neural networks*, 1, 251-265.

Narendra, K.S. and Parthasarathy, K. (1990) Identification and control of dynamical systems using neural networks, *IEEE Trans. on Neural Networks*, **1**(1), 4-27.

Narendra, K.S. and Parthasarathy, K. (1991) Gradient methods for the optimization of dynamical systems containing neural networks, *IEEE Trans. on Neural Networks*, **2**(2), 252-262.

Pham, D.T. and Liu, X. (1990) State space identification of dynamic systems using neural networks, *Engineering Applications of Artificial Intelligence*, **3**, 198-203.

Psaltis, D., Sideris, A. and Yamamura, A. A. (1988) A multilayered neural network controller, *IEEE Control Systems Magazine*, 17- 21.

Rovithakis, G.A. and Christodoulou, M.A. (1994) A robust direct adaptive regulation architecture using dynamic neural network models, *Int. Conf. on System, Man and Cybernetics*, San Antonio, Texas, USA, October 1994, 1110-1115.

Rumelhart, D. E. and McClelland, J. L. (1986) *Parallel Distributed Processing: explorations in the microstructure of cognition*, vol.1: Foundations, MIT Press.

Saint-Donat, J., Bhat, N. and McAvoy, T.J. (1991) Neural net based model predictive control, *Int. J. Control*, **54**(6), 1453-1468.

Walker, M., Haster, P. and Akers, L. (1989) A CMOS neural network for pattern association, *IEEE MICRO*, **9**(5), 68-74.

Wang, H., Brown, M., Harris, C.J. (1994) Fault detection for a class of unknown non-linear systems via associative memory networks, *Proc. IMechE, Part I, J. of Systems and Control Engineering*, **208**(12), 101-107.

Warwick, K., Irwin, G.W., Hunt, K.J. (1992) *Neural Networks for Control and Systems*, London: Peregrinus.

Werbos, P.J. (1991) An overview of neural networks for control, *IEEE Control Systems Magazine*, 40-41.

Werbos, P.J. (1991) Neural networks for modelling and optimization over time, *Eighth Int. Conf. on Math. and Computer Modelling*, Maryland, USA, 6.

Xu, L., Jiang, J.P. and Zhu, J. (1994) Supervised learning control of a nonlinear polymerisation reactor using the CMAC neural network for knowledge storage, *Proc. IEE, Part D*, 141(1), 33-38.

Yamada, T. and Yabuta, T. (1990) An extension of neural network direct controller, *IEEE International Workshop on Intelligent Robots and Systems*, July, Tsuchiura, Japan.

Chapter 7 Neuromorphic Fuzzy Controller Design

This chapter shows that a single-input single-output (SISO) Fuzzy Logic Controller (FLC) [Mamdani, 1974; Lee, 1990a, 1990b] can be modelled as a neural network which can be trained using a Genetic Algorithm (GA). The GA is employed to determine the membership functions for the input variable, the quantisation levels of the output variable and the elements of the relation matrix of the FLC. The reasons for such a neuromorphic FLC are provided. The structure of the NN model proposed for an FLC and its GA-based training procedure are explained. Results for the simulated control of a time-delayed linear second-order plant and a non-linear plant are also given. In this chapter, it is assumed that the reader is familiar with fuzzy logic control and genetic algorithms. For a basic introduction to these topics, see Appendix B and Appendix C.

7.1 Integrating Neural Networks and FLCs

There are two problems with the design of a FLC. The first is the selection of membership functions for the input and output variables, which are generally determined intuitively by human experts. The second is the determination of a relation matrix representing the mapping between the input and output spaces. These problems are due to the lack of a learning ability in FLC systems [Pham and Karaboga, 1991].

Neural networks (NN) provide a different approach to problem solving from linguistic or algorithmic systems such as FLCs. As discussed in previous chapters, they have two main features which are their noise-resistant parallel distributed structure and their ability to learn from examples. NNs have a wide spectrum of actual and potential applications ranging from pattern recognition and prediction to dynamic system modelling and control [Maren et al,1990; Pham and Liu,1991].

7.1.1 Representation of a Simple Fuzzy Logic Controller as a Neural Network

A SISO FLC represented as a NN system is depicted in Figure 7.1.

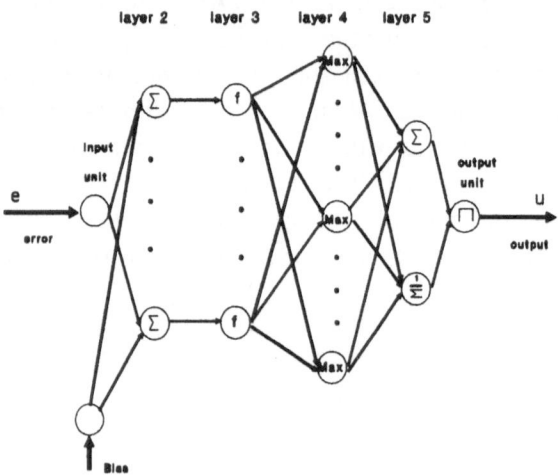

Figure 7.1 Neural network representation of a FLC

In the Figure, e is the "crisp" error between the actual and desired outputs of the plant being controlled. This error is distributed by the input neuron, the only neuron in layer 1, to the neurons in the next layer (layer 2). Before the distribution process, it is scaled into a predetermined range by being multiplied with a positive factor (c_1)

representing the input mapping coefficient of the FLC. The actual value fed to each neuron in layer 2 is thus :

$$a_1 = c_1 e \tag{7.1}$$

where c_1 represents the weights of the links between the input layer and layer 2.

The input of a neuron in layer 2 can be expressed as :

$$I_{2i} = a_1 + b_i \qquad i = 1...r \tag{7.2}$$

where bi is the bias to neuron i in layer 2. The activation functions of all neurons in this layer are linear and thus their outputs a2i are

$$a_{2i} = I_{2i} \qquad i = 1...r \tag{7.3}$$

Layers 1, 2 and 3 together implement membership functions which map a continuous domain into the predetermined range [0,1]. In this representation, two different types of membership functions are used. The first is bell-shaped and the second is sigmoidal. The general expressions for these functions are :

$$f(x) = exp(-(x-z)^2 t^2) \qquad \text{(bell-shaped function)} \tag{7.4}$$

$$f(x) = 1/(1+exp(-(x-z)t)) \qquad \text{if x >= 0 (sigmoidal function)} \tag{7.5}$$
$$= 1/(1+exp((x-z)t)) \qquad \text{if x < 0}$$

Variable x in Equations (7.4) and (7.5) is equal to a_1 in Equation (7.1). The biases b_i of the neurons in layer 2 determine the positions of the "centres" (z) of the membership functions. Since these are fixed for the two extreme regions and the middle region of the input domain, the corresponding biases $(b_1, b_r$ and $b_{(1+r)/2})$ are also fixed.

Layer 3 has the same number of neurons (r) as layer 2. The neurons in the two layers are linked one-to-one. The variable weights w_{2i} of the connections between layer 2 and layer 3 define the "widths" (t) of the membership functions. The output of each neuron in layer 3 has a value in the interval [0,1]. The output vector $A_3 = (a_{31}, a_{32},..., a_{3r})$ represents the fuzzy set obtained from the crisp input to the network. The activation functions of the first and last units in this layer are sigmoidal and their outputs a_{31} and a_{3r} are :

$$a_{31} = 1 / (1 + exp(I_{31})) \qquad\qquad (7.6)$$

$$a_{3r} = 1/(1 + exp(-I_{3r})) \qquad\qquad (7.7)$$

where $I_{31} = w_{21} a_{21}$ and $I_{3r} = w_{2r} a_{2r}$.

The activation functions of the other neurons in layer 3 are bell shaped and their outputs a_{3i} are :

$$a_{3i} = exp(-(I_{3i})^2) \qquad\qquad i = 2... \; r-1 \qquad\qquad (7.8)$$

where $I_{3i} = w_{2i} a_{2i}$.

Layer 4 and its links with layer 3 implement the Max-Product operator for fuzzy inferencing. The connections from layer 3 to layer 4 have modifiable weights w_{3i} and carry out the Product operation.

The neurons in layer 4 perform the Max operation. The output vector of this layer represents a fuzzy set. The components a_{4i} of that vector are :

$$a_{4i} = I_{4i} = \overset{r}{\underset{i=1}{Max}} (w_{3i} \, a_{3i}) \qquad\qquad (7.9)$$

where a_{4i} is the membership value of element i of the output fuzzy set. Element i represents the ith output quantisation level which is yet to be determined.

Layer 5 has two units. The first unit produces the weighted sum of the values of all output quantisation levels. The second adds up the weights used to compute the weighted sum, i.e. the outputs of layer 4, and then inverts the result.

The total inputs I_{51} and I_{52} to both units are given by :

$$I_{51} \; or \; I_{52} = \sum_{i=1}^{s} w_{4i} \, a_{4i} \qquad\qquad (7.10)$$

where s is the number of neurons in layer 4. For the first unit, w_{4i} defines the value of the ith output quantisation level and is obtained through training. For the second unit, all w_{4i}'s (i = 1 to s) are fixed at unity. The outputs of these units are :

$$a_{51} = I_{51} \qquad\qquad (7.11)$$

$$a_{52} = 1/I_{52} \qquad\qquad (7.12)$$

The outputs a_{51} and a_{52} of layer 5 are sent via links with unit weights to the output neuron. The latter is a product unit which multiplies a_{51} and a_{52} together to yield its net input I_6 :

$$I_6 = a_{51}a_{52} \qquad\qquad (7.13)$$

To obtain the final output u of the network, the net input I_6 is multiplied with a factor c_2 representing the output scaling coefficient of the FLC :

$$u = a_6 = c_2 I_6 \qquad\qquad (7.14)$$

7.1.2 Learning algorithm

A learning algorithm is used to search for the values of the variable weights in the neural network. The search space is complex as there are restrictions on the values of some weights and the network contains processing elements such as the Max units and the Product unit. The genetic algorithm (GA) [Holland,1975; Goldberg, 1989; Davis,1991], which is an efficient search procedure for large and complicated search spaces, has been selected to implement the neural network learning algorithm. The flow chart of the GA used is shown in Figure 7.2. From that figure, the first step in a GA is initialisation of the string population. As explained in Appendix C, this procedure randomly generates a set of strings for the GA to operate upon. Each binary string in the string population represents a potential solution, that is the values for the set of modifiable weights of the NN. A string has the following form :

$$\eta_1 \eta_2 \cdots \eta_i \cdots \eta_p \qquad\qquad (7.15)$$

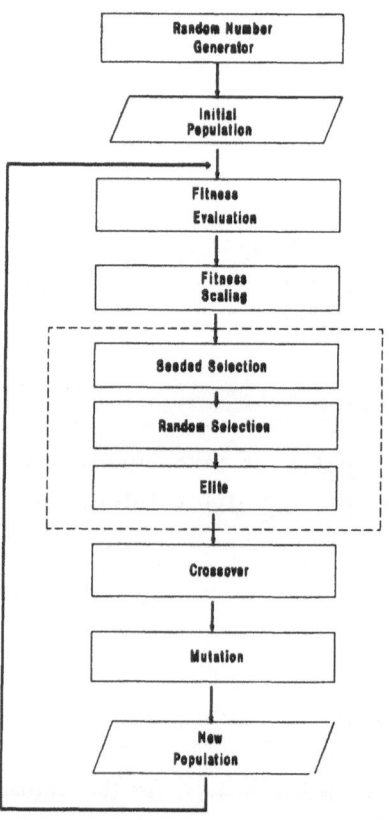

Figure 7.2 Flowchart of the used standard genetic algorithm

where η_i is the value of modifiable weight w_i and is itself a binary string. The value of p depends on the number of fuzzy values of the fuzzy input variable (r) and the number of quantisation levels for the output space (s).

In the evaluation procedure, NNs represented by the strings in the population are reconstituted and their performances (fitnesses) are computed. A fitness scaling procedure is adopted prior to the performance evaluation to enable a finer discrimination between good strings and even better ones. The performance of a string is measured according to the accumulated error in the step response of the plant. The ITAE formula is again used to compute this error.

7.2 Results of Neuromorphic Fuzzy Controllers Design

Before training the network its overall structure has to be set up. The number of fuzzy values (fuzzy sets covering the input space) and the number of quantisation levels into which the output space is divided has to be decided since they determine the number of units in layers 2, 3 and 4, respectively. The type of membership functions and the upper and lower limits of their ranges also have to be defined before the training.

Seven fuzzy partitions are used for the input space and thus the numbers of neurons in layers 2 and 3 are seven. The output space is divided into eleven levels giving eleven neurons in layer 4. Therefore, the total number of variable weights is 99, of which 11 are for membership functions, 11 for the output quantisation levels and 77 for the elements of the relation matrix. Each weight is represented by 8 bits.

The GA is run for 400 generations. The population size is 100. The other parameters of the GA are:

Crossover rate: 0.85
Mutation rate : 0.01
Generation gap: 0.95

The model proposed is tested on two different plants: a linear and a non-linear plant.

7.2.1 Plant 1

The first plant to be controlled is a time-delayed second-order system described by Equation 7.16.

$$G(s) = (K \, e^{-\tau s})/(a \, s + b)^2 \qquad (7.16)$$

It is assumed that a sampled data system with a zero-order hold (ZOH) is adopted. The sampling period T is selected as 0.1 sec. The plant characteristic parameters are:

$$\tau = 0.4, \quad a = 0.3, \quad b = 1.0, \quad K = 1.0 \tag{7.17}$$

The transient response of the plant under the control of the trained FLC is shown in Figure 7.3. The membership functions and the values of the quantisation levels obtained for the input space and the output space are given in Figure 7.4 and Table 7.1, respectively.

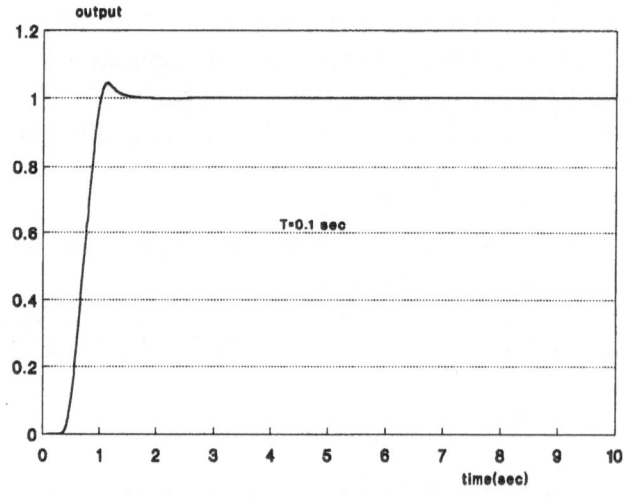

Figure 7.3 Transient response of linear plant controlled by the GA trained FLC

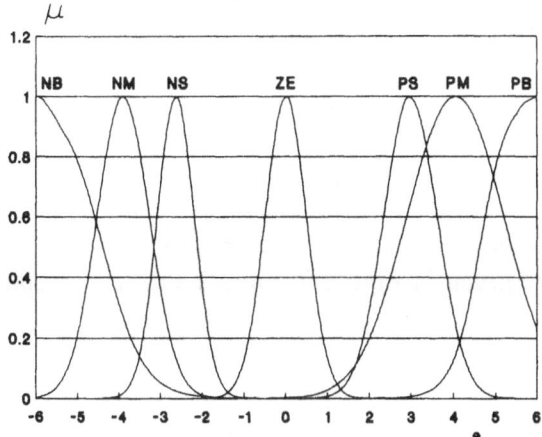

Figure 7.4 Membership functions of the FLC
trained by GA (Linear plant)

Table 7.1 Values of output quantisation levels
obtained for the linear plant by the GA

Quantisation levels	Value
1	-4.716
2	-4.209
3	-2.755
4	-2.139
5	-0.833
6	-0.029
7	1.257
8	2.378
9	3.441
10	4.127
11	5.316

7.2.2 Plant 2

The second plant used in the simulation is a non-linear plant with the following discrete-time equation :

$$y(k) = y(k-1) \, / \, (1.5 + y^2(k-1)) - 0.3 \, y(k-2) + 0.5 \, u(k-1) \qquad (7.18)$$

The transient response of the plant under the control of the trained FLC is shown in Figure 7.5. The membership functions and the values of the quantisation levels obtained for the input space and the output space are given in Figure 7.6 and Table 7.2, respectively.

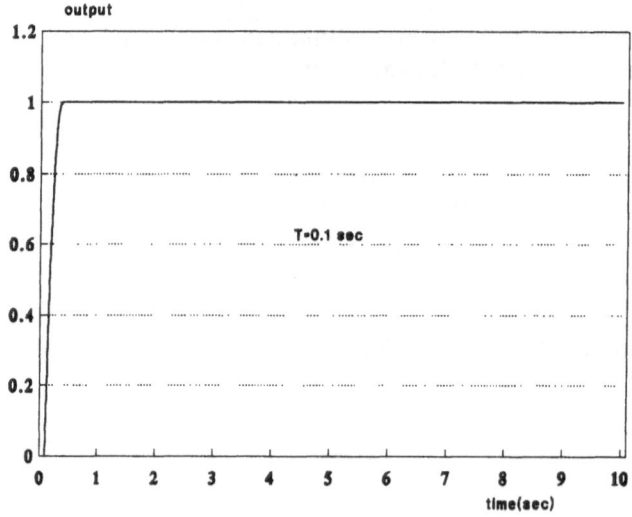

Figure 7.5 Response of nonlinear plant controlled by GA trained FLC

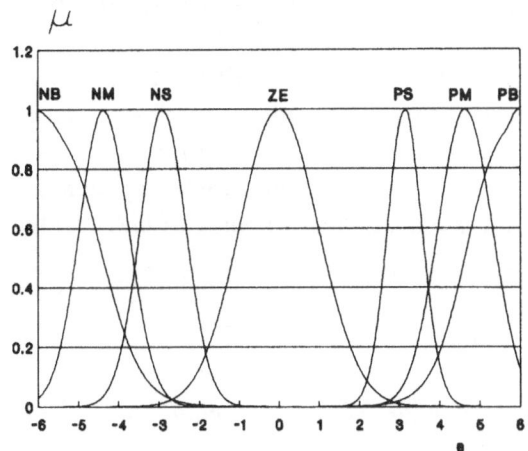

Figure 7.6 Membership functions of FLC trained by GA
(nonlinear plant)

Table 7.2 Values of output quantisation levels obtained
for the nonlinear plant by the GA

Qualisation levels	Value
1	-4.873
2	-3.873
3	-2.676
4	-1.598
5	-0.559
6	0.127
7	0.676
8	2.108
9	3.300
10	4.225
11	5.363

7.3 Summary

A model for SISO FLCs based on a feedforward NN structure has been presented. An advantage of modelling FLCs as NNs is that it facilitates a uniform treatment of the FLC parameters, enabling all of them to be obtained through training. A GA is employed to train an FLC as if it were an ordinary NN. The results for a time-delayed second-order linear system and a non-linear system are presented, showing the fast and accurate performance of GA-trained FLCs.

References

Davis, L. (1991) *Handbook of genetic algorithms*, New York: Van Nostrand Reinhold.

Goldberg, D.E. (1989) *Genetic algorithms in search, optimization, and machine learning*, Reading, MA: Addison-Wesley.

Holland, J.H. (1975) *Adaptation in natural and artificial systems*, Ann Arbor, MI: The university of Michigan Press.

Lee, C.C. (1990a) Fuzzy logic in control systems: Fuzzy logic controller, Part I, *IEEE Trans. on Systems, Man and Cybernetics*, 20(2), 404-418.

Lee, C.C. (1990b) Fuzzy logic in control systems: Fuzzy logic controller, Part II, *IEEE Trans. on Systems, Man, and Cybernetics*, 20(2), 419-435.

Mamdani, E.H. (1974) Applications of fuzzy algorithms for control of simple dynamic plant, *Proc. IEE*, 121(12), 1585-1588.

Maren, A., Harston, C. and Pap, R. (1990) *Handbook of Neural Computing Applications*, London: Academic Press.

Pham, D.T. and Karaboga, D. (1991) A new method to obtain the relation matrix of fuzzy logic controllers, *Proc. Sixth Int. Conf. on Artificial Intelligence in Engineering*, Oxford, 567-581.

Pham, D.T. and Liu, X. (1991) Neural networks for discrete dynamic system identification, *J. of Systems Engineering*, 1(1), 51-60.

Chapter 8 Robot Manipulator Control Using Neural Networks

The control of a multi-input-multi-output (MIMO) plant is a difficult problem when the plant is nonlinear and time-varying and there are dynamic interactions between the plant variables. A good example of such a plant is an articulated robot with two or more joints handling a changeable load.

Conventional methods of designing controllers for a MIMO plant like a multi-joint robot generally require, as a minimum, knowledge of the structure of an accurate mathematical model of the plant [Spong and Vidyasagar, 1989; Tomizuka et al, 1988]. In many cases, the values of the parameters of the model also need to be precisely known [Craig et al, 1987; Slotine et al, 1987; Khosla, 1986].

Neural networks, which can learn the forward and inverse dynamic behaviours of complex plants on-line, offer alternative methods of realising MIMO controllers capable of adapting to environmental changes. In theory, the design of a neural-network-based control system should be relatively straightforward as it does not require any prior knowledge about the plant. However, practical problems regarding the neural network structure to be adopted, the number of input units and the training procedure including training signal pattern would still need to be addressed.

Neural control approaches involve using neural networks to learn the modelling and control task automatically [Miller et al, 1990; Goldberg and Pearlmutter, 1988]. A neural controller invariably includes a neural network that has been trained to model in some way the inverse dynamics of the plant. In the schemes proposed to-date, this training can be carried out by adopting: (i) a direct (or generalised) learning architecture or (ii) an indirect (or specialised) learning arrangement.

Direct learning being separate from control is generally implemented off-line, and indirect learning is carried out on-line with the plant under the control of the "trainee" neural network. The main problems with these methods of obtaining a neural-network-based inverse controller have been described in [Oh, 1993; Psaltis et al, 1989; Colombano et al, 1991].

In this chapter, a new approach to the neural control of a multi-joint robot is described. Three neural networks are employed in total, the first to learn the dynamics of the robot, the second to model its inverse dynamics and the third, a copy of the second neural network, to control the robot. In addition to using on-line direct learning which is a more efficient way to evolve an adaptive neural controller, the approach also differs from previous work by other authors [Goldberg and Pearlmutter, 1988; Miller et al, 1987; Kawato et al, 1988; Ciliz, 1990; Kung and Hwang, 1991; Ozaki et al, 1991] on one important aspect: it is based on input-output identification which is simpler to implement than the state-space identification approach adopted hitherto.

8.1 Modelling of a Multi-joint Robot

Figure 8.1(a) shows a general block diagram for a robot with n driven joints as an example of a MIMO plant.

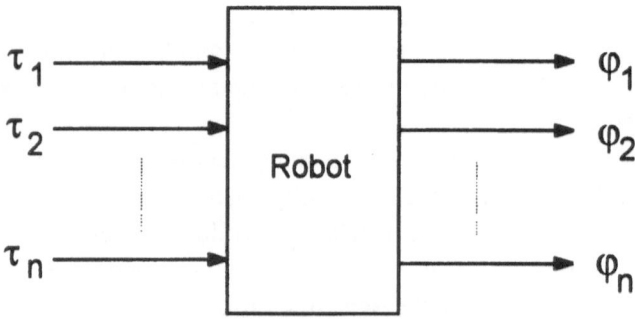

Figure 8.1 (a) Block diagrams for an n-joint robot: Open-loop robot.

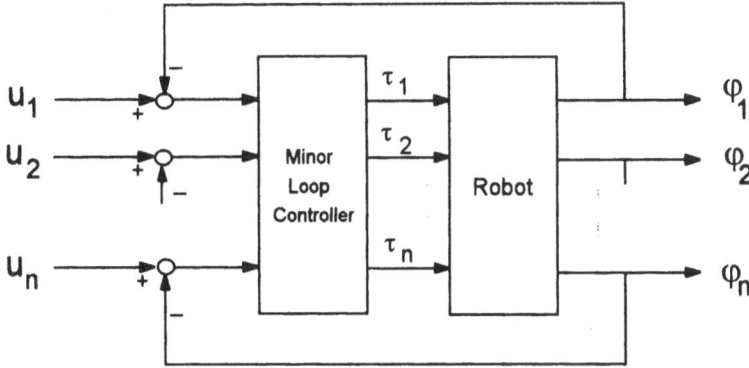

Figure 8.1 (b) Block diagrams for an n-joint robot:
Robot with a minor-loop controller

The inputs to the plant are torques $\tau = [\tau_1 \ \tau_2 \ ... \ \tau_n]^T$ and the outputs are actuator rotation angles $\varphi = [\varphi_1 \ \varphi_2 \ ... \ \varphi_n]^T$. The torque commands are normally issued by an overall plant controller. However, often because of practical considerations, such as the integration characteristic of actuators, high lags and low system gains due to large inertias, and nonlinearities due to friction, backlash and dynamic coupling, it is necessary to add a minor-loop controller to simplify the design of the overall controller and obtain more stable control [Spong and Vidyasagar, 1989; D'Souza, 1988; Kuo, 1982]. Figure 8.1(b) depicts a system with a minor-loop controller. The inputs to the system are now $\mathbf{u} = [u_1 \ u_2 \ ... \ u_n]^T$ where $u_1 \ u_2 \ ... \ u_n$ are the target angles for joints 1,2,..., n.

To construct a neural control system for the plant shown in Figures 8.1(a) and 8.1(b) requires neural models of both the forward and inverse dynamics of the plants. As will be seen later, the forward dynamics model, which has the same inputs (τ or $\mathbf{u}$) and outputs (φ) as the plant, is used in a predictive mode to compute the next outputs of the plant and estimate future deviations from the desired outputs. The inverse dynamics model, which has φ as its inputs and τ or $\mathbf{u}$ as its outputs, is employed essentially to provide control signals to the plant.

It has been demonstrated, respectively, that the modified Jordan network can readily be configured to model the forward dynamics [Pham and Oh, 1992a] and inverse dynamics [Pham and Oh, 1992b] of a plant. Figures 8.2(a) and 8.2(b) are schematic diagrams of the modified Jordan networks for forward and inverse dynamics modelling.

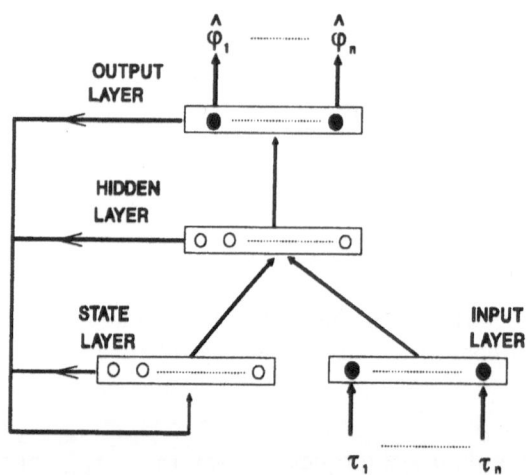

Figure 8.2 (a)   Modified Jordan network for forward dynamics modelling

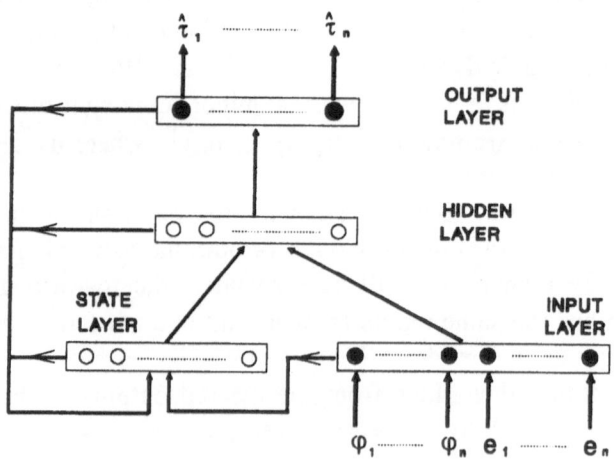

Figure 8.2 (b) Modified Jordan networks for inverse dynamics modelling

The modification to the basic Jordan network consists of connecting the outputs of neurons in the hidden layer back to the inputs of the corresponding neurons in the state layer. This modification allows the network to be trained to represent arbitrary dynamic systems. Note that, in addition to inputs φ, the inverse dynamics modelling network also has inputs $\mathbf{e} = [e_1 \; e_2 \; ... \; e_n]^T$ which are the errors between the desired and actual actuator rotation angles. These error terms are needed to achieve closed-loop control.

8.2 Control System

The control problem is how to track an arbitrary reference trajectory under the following conditions: time-varying nonlinear and multivariable plant with dynamic coupling between the variables, unknown plant model structure, and input-output data captured from the plant at discrete sampling intervals being the only information available about the plant.

As shown in Figure 8.3, the control system adopted comprises a feedforward controller ($\hat{\Phi}^c$), which is a copy of the neural network $\hat{\Phi}$ used to learn the inverse dynamics of the plant, a feedback controller and an error predictor. The latter uses a model of the plant obtained by the forward dynamics modelling network $\hat{\Psi}$. Both $\hat{\Phi}$ and $\hat{\Psi}$ are trained on-line during control to give the system the ability to adapt to change. The control system is an extended version of that described in [Oh, 1993], except that an outer feedback controller (a feedback servo controller) is incorporated within this control system, it is a MIMO system and it does not have an adaptation critic mechanism. The critic mechanism is omitted to reduce complexity. Its omission is possible because of the presence of the outer feedback controller.

From Figure 8.3, the control input to the robot is given by:

$$\mathbf{u}_c = \hat{\Phi}_c + \mathbf{K}_f \, \mathbf{e} \tag{8.1}$$

where $\hat{\Phi}_c$ is the output of $\hat{\Phi}^c$ and $\mathbf{K}_f$, the gain of the feedback controller.

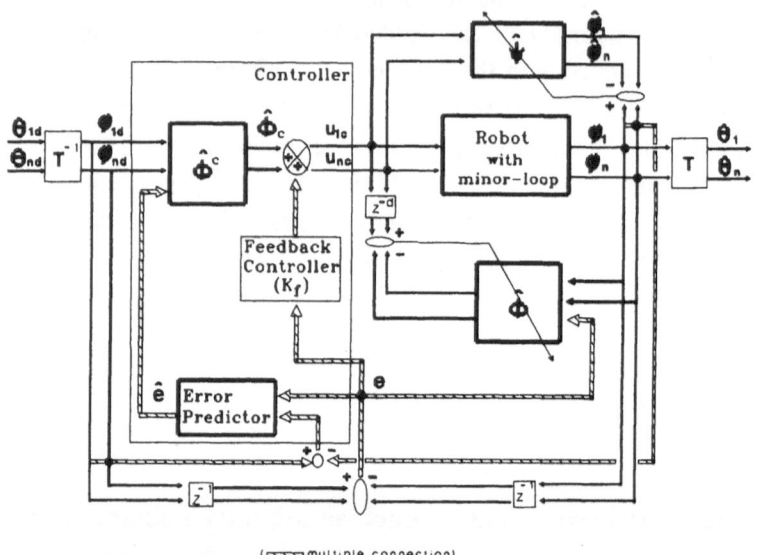

Figure 8.3 Block diagram of control system

Due to the recurrent nature of the adopted neural network, $\hat{\Phi}_c$ is a function of the explicit inputs to $\hat{\Phi}^c$ (φ_d and $\hat{e}$) as well as the implicit inputs (past values of φ_d, $\hat{e}$, and $\hat{\Phi}_c$), that is,

$$\hat{\Phi}_c(k) = \hat{\Phi}^c\{\varphi_d(k+1),\ \hat{e}(k+1),\ \overset{*}{\varphi_d},\ \overset{*}{\hat{e}},\ \overset{*}{\hat{\Phi}_c}\} \qquad (8.2)$$

where $\overset{*}{\varphi_d}$, $\overset{*}{\hat{e}}$, $\overset{*}{\hat{\Phi}_c}$ are past dynamic memory values of φ_d, $\hat{e}$ and $\hat{\Phi}_c$ respectively.

In equation (8.2), $\hat{e}(k+1)$ is the predicted total error obtained from the error predictor as follows:

$$\hat{e}(k+1) = K_1 e(k) + K_2 [\varphi_d(k+1) - \hat{\varphi}(k+1)] \qquad (8.3)$$

where K_1 and K_2 are the gains of the predictor, $e(k)$, the actual control error at instant k and $[\varphi_d(k+1) - \hat{\varphi}(k+1)]$, a predicted error term for instant (k+1). $\hat{\varphi}(k+1)$ is produced by the forward dynamics modelling network $\hat{\Psi}$ thus:

$$\hat{\varphi}(k+1) = \hat{\Psi}\{u_c(k), \ u_c^*, \ \hat{\varphi}^*\} \tag{8.4}$$

In equation (8.4), u_c^* and $\hat{\varphi}^*$ are past dynamic memory values of u_c and $\hat{\varphi}$.

8.3 Application to a Two-joint Robot Arm

This section describes the forward and inverse dynamics modelling and control of a two-degree-of-freedom robot arm.

Figure 8.4 shows the arm which is a planar device of the Scara configuration [Craig, 1987; Makino et al, 1980], comprising two main links with two actuated joints. Actuator 1 applies torque τ_1 to drive joint 1 which directly connects link 1 (with inertia I_1) to the base of the arm. Actuator 2 applies torque τ_2, via a band and pulley system (with inertia I_2), to drive joint 2 connecting link 2 (with inertia I_3 and mass m_2) to link 1. The distance between the axes of joints 1 and 2 is denoted l_1 and the distance between the centre of mass of link 2 and the axis of joint 2 is termed l_2. The angles of rotation of the actuators are φ_1 and φ_2 respectively.

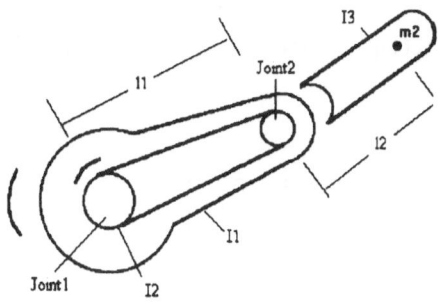

Figure 8.4 (a) Top view of Scara robot arm.

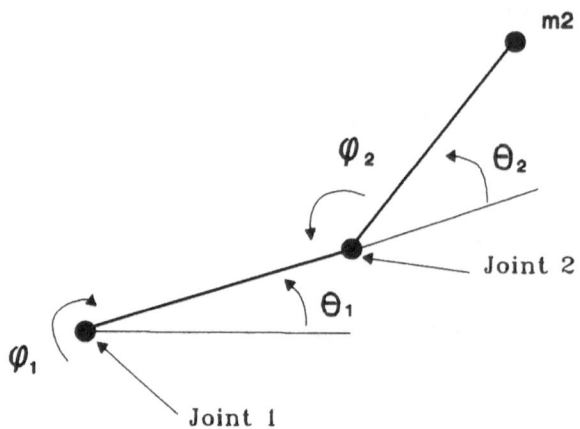

Figure 8.4 (b) Actuator and joint rotation angles of Scara robot arm

The equations of motion of the arm can be expressed in matrix form as follows [Craig, 1987]:

$$
\begin{bmatrix} A_1 & -A_3 \cos(k_h(\varphi_1+\varphi_2)) \\ -A_3 \cos(k_h(\varphi_1+\varphi_2)) & A_2 \end{bmatrix} \begin{bmatrix} \ddot{\varphi}_1 \\ \ddot{\varphi}_2 \end{bmatrix}
$$

$$
= \begin{bmatrix} 0 & A_3 k_h \sin(k_h(\varphi_1+\varphi_2))\dot{\varphi}_2 \\ A_3 k_h \sin(k_h(\varphi_1+\varphi_2))\dot{\varphi}_1 & 0 \end{bmatrix} \begin{bmatrix} \dot{\varphi}_1 \\ \dot{\varphi}_2 \end{bmatrix}
$$

$$
- \begin{bmatrix} A_4 \dot{\varphi}_1 + A_6 \operatorname{sgn}(\dot{\varphi}_1) \\ A_5 \dot{\varphi}_2 + A_7 \operatorname{sgn}(\dot{\varphi}_2) \end{bmatrix} + \begin{bmatrix} \tau_1 \\ \tau_2 \end{bmatrix} \qquad (8.5)
$$

where

$$
A_1 = I_1 + m_2 l_1^2
$$
$$
A_2 = I_2 + I_3 + m_2 l_2^2
$$

$A_3 = m_2 \, l_1 \, l_2$

A_4, A_5 = viscous friction coefficients

A_6, A_7 = Coulomb friction torques

k_h = actuator encoder constant

In the modelling and control experiments described below, the kinematic and inertial parameters of the arm are chosen as: l_1=0.4 m, l_2=0.35 m, m_2=10 Kg, I_1=1.8 Kg m^2, I_2=0.041 Kg m^2 and I_3=0.134 Kg m^2. Thus A_1=3.4 Kg m^2, A_2=1.4 Kg m^2, and A_3=1.4 Kg m^2. The friction parameters are set at A_4=5.0 Kg m^2/s, A_5=2.0 Kg m^2/s, and A_6=A_7=0.0. The actuator encoder constant k_h is equal to 0.5.

Given the external torques τ_1 and τ_2, the actual rotation angles φ_1 and φ_2 can be obtained from equation (8.5) by applying a numerical analysis technique, such as the Runge-Kutta 4th-order method [Gerald and Wheatley, 1989]. In this work, a Runge-Kutta subroutine written in C is employed. The state variables are $[\varphi_1, \dot{\varphi}_1, \varphi_2, \dot{\varphi}_2]$ and the time step is 0.01 second.

The joint angles, $\theta = [\theta_1 \ \theta_2]^T$, can be obtained from the actuator rotation angles, $\varphi = [\varphi_1 \ \varphi_2]^T$, using the following equation:

$$\theta = \mathbf{T} \, \varphi \tag{8.6}$$

where the transformation matrix $\mathbf{T}$ is equal to $\begin{bmatrix} -k_h & 0 \\ k_h & k_h \end{bmatrix}$.

The forward dynamics modelling and inverse dynamics modelling of the robot are carried out using the modified Jordan networks shown in Figures 8.5(a) and 8.5(b) respectively, which are also extended version of those described in [Oh, 1993]. The structural and training parameters for the networks are as shown in Table 8.1. The delay term z^{-d} in the inverse modelling scheme is equal to z^{-1} [Pham and Oh, 1992].

To generate training signals for the neural networks when the robot does not have a minor feedback loop, a controller has to be employed which enables the setting of arbitrary target joint rotation angles. The chosen controller is a proportional system with gain $\mathbf{K_b} = \begin{bmatrix} 20 & 0 \\ 0 & 20 \end{bmatrix}$.

Note that this controller is not used after the training set has been constructed.

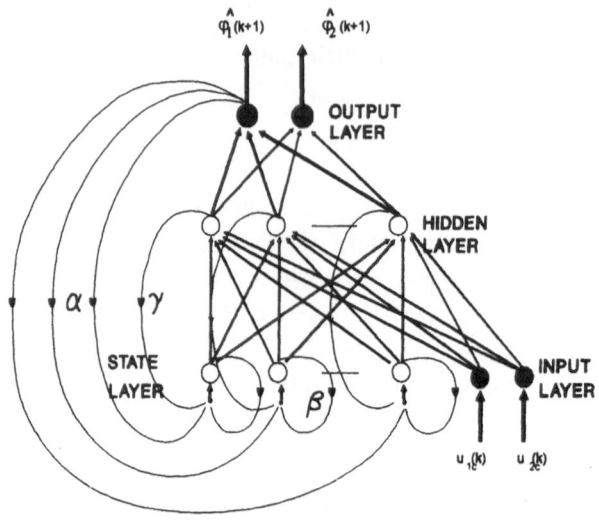

Figure 8.5 (a) Neural network models of the robot: forward dynamics.

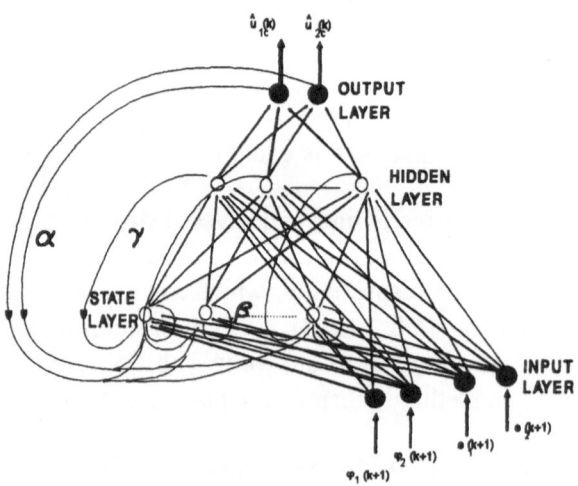

Figure 8.5 (b) Neural network models of the robot: inverse dynamics.

In the case where a minor loop is employed, a proportional feedback controller is also adopted as the minor controller during the learning and control processes. The gain of that controller is $\mathbf{K_m} = \begin{bmatrix} 100 & 0 \\ 0 & 100 \end{bmatrix}$.

Figures 8.6 and 8.7 show the forward and inverse dynamics responses of the robot and the neural network models for the case without the minor feedback loop. Figures 8.8 and 8.9 illustrate the responses obtained with the minor feedback loop. It can be seen that in both cases the neural network responses are virtually indistinguishable from those of the robot. However, training of neural networks is greatly facilitated when the minor loop is incorporated [Pham and Oh, 1992]. See the number of training iterations required for the different cases in Table 8.1. The pattern training method is used. Each training iteration consists of presenting an input data item to the neural networks and modifying the weights of the networks to bring their outputs closer to the desired outputs. The input-output data items for the neural networks are taken directly from the plant.

The neural networks trained with the minor loop are used to control the robot according to the scheme depicted in Figure 8.3. The gain of the outer feedback controller $\mathbf{K_f}$ is set to $\mathbf{K_f} = \begin{bmatrix} 20 & 0 \\ 0 & 20 \end{bmatrix}$. The gains of the error predictor (see eq.(8.3)) are $\mathbf{K_1} = \begin{bmatrix} 1 & 0 \\ 0 & 1 \end{bmatrix}$ and $\mathbf{K_2} = \begin{bmatrix} 1 & 0 \\ 0 & 1 \end{bmatrix}$.

The system is given the following target joint rotations: $\theta_{1d}=\pi/12.0\sin(2\pi*0.25t)$ and $\theta_{2d}=\pi/12.0\sin(2\pi*0.25t)$. Figure 8.10 plots these target rotations and the actual rotations obtained. After an initial settling period, the target and actual rotations almost coincide with one another. Figure 8.11 shows the responses of the robot shortly after the targets are changed to $\theta_{1d}=1.5*\pi/12.0\sin(2\pi*0.25t)+1.5*\pi/24.0\sin(2\pi*0.125t)$ and $\theta_{2d}=\pi/12.0\sin(2\pi*0.25t)$. This shifts the operating ranges for θ_1 and θ_2 outside the ranges in which the neural networks have been trained and disturbs the ability of the nonlinear control system to follow targets. However once the neural networks have adapted to the new operating range, which requires some 6,000 on-line training iterations, the difference between the target and actual outputs is reduced. Figure 8.12 shows the responses of the robot after adapting to the new operating range.

Table 8.1 Control parameters with trained inverse controller

Parameter		η	μ	β	n	T	Hid. layer activation
Plant without minor loop	Forward	0.005	0.001	0.8	12	300,000	Hyperbolic tangent
	Inverse	0.01	0.001	0.8	10	300,000	Hyperbolic tangent
Plant with minor loop	Forward	0.005	0.001	0.8	12	100,000	Hyperbolic tangent
	Inverse	0.05	0.001	0.8	10	100,000	Hyperbolic tangent

(η: learning coefficient; μ: momentum term; β: self feedback connection in hidden layer; n: number of hidden neurons; T: training iterations)

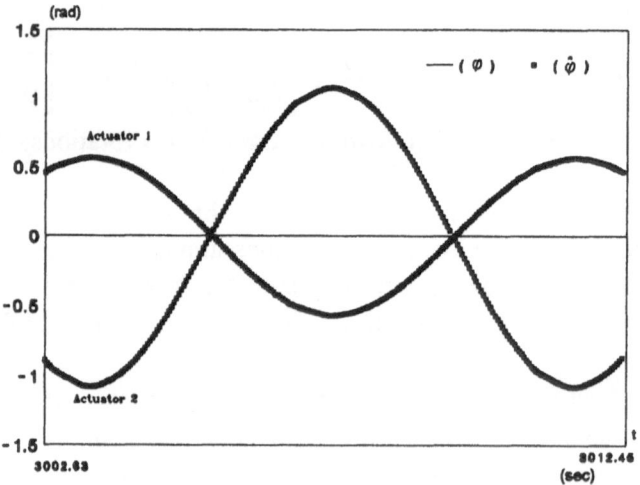

Figure 8.6 Responses of plant and forward dynamics model (without minor loop)

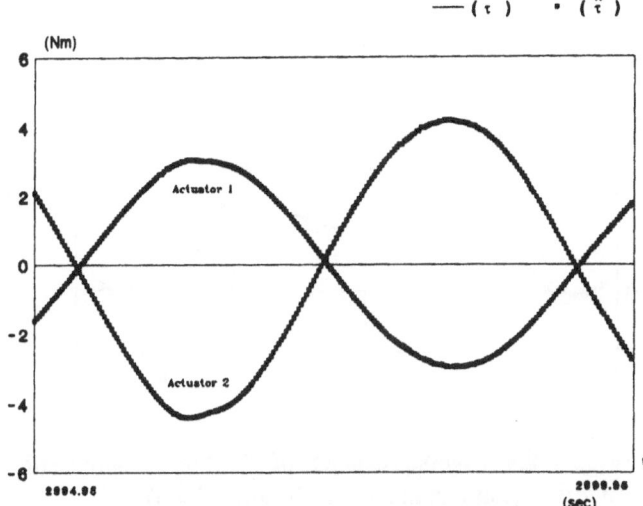

Figure 8.7 Inverse response of plant and response of inverse dynamics model(without minor loop)

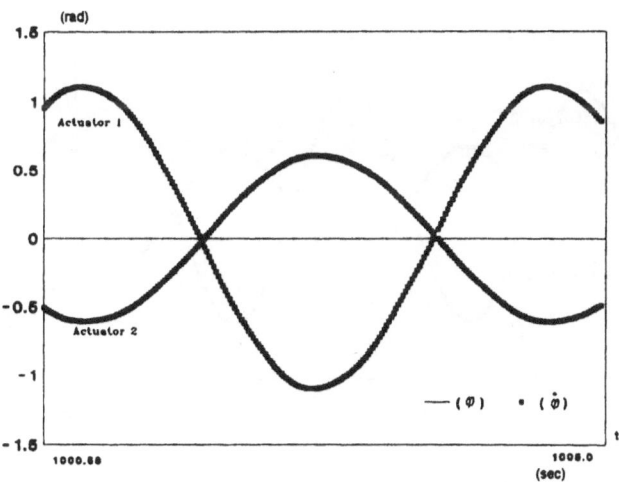

Figure 8.8 Responses of plant and forward dynamics model (with minor loop)

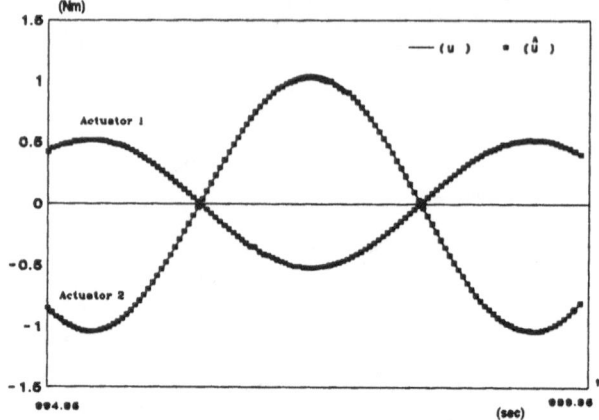

Figure 8.9 Inverse response of plant and response of inverse dynamics model (with minor loop)

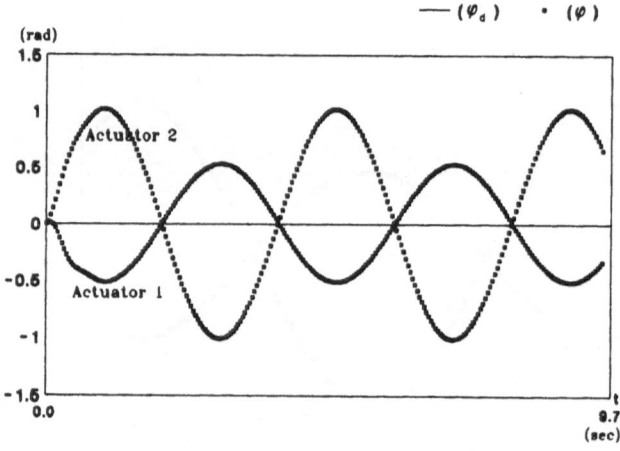

Figure 8.10 Control responses with trained inverse controller

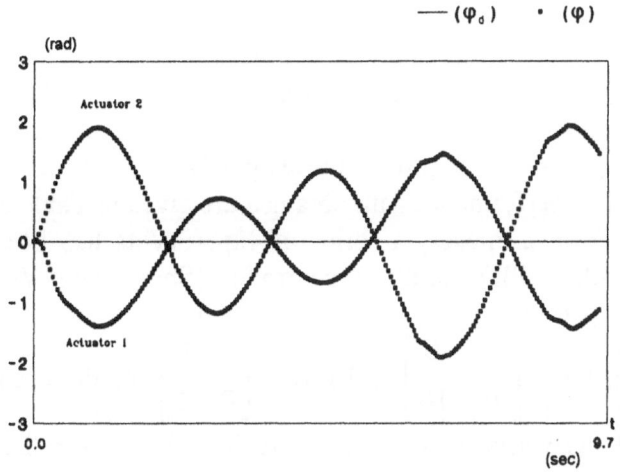

Figure 8.11 Control responses for a new operation range (before adaptation)

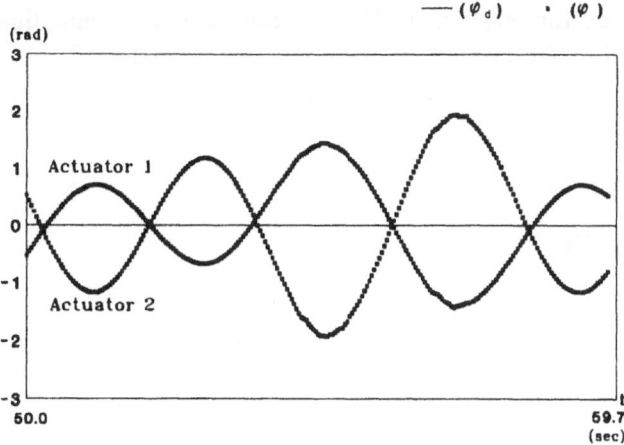

Figure 8.12 Control responses for a new operating range (after adaptation)

The adaptive ability of the neural controller is also tested on a problem where there is a change in the dynamic parameters of the robot. It is assumed that mass m_2 is increased from 10 Kg to 20 Kg. The dynamic parameters A_1, A_2, and A_3 become 5.0 Kg m^2, 2.725 Kg m^2, and 2.8 Kg m^2 respectively. An input signal comprising a sinusoidal part, a ramp and a step is applied. Figure 8.13 plots the desired and actual responses before the parameter change. Figures 8.14 and 8.15 show the responses before and after adapting to the change (the adaptation process also requires approximately 6,000 on-line training iterations). The performance data obtained are given in Table 8.2.

For comparison purposes, a series of experiments has been further conducted with a PID feedback controller [Spong and Vidyasagar, 1989; Craig, 1989]. The PID controller gains are chosen as $K_P = \begin{bmatrix} 100 & 0 \\ 0 & 100 \end{bmatrix}$, $K_I = \begin{bmatrix} 10 & 0 \\ 0 & 10 \end{bmatrix}$ and $K_D = \begin{bmatrix} 1 & 0 \\ 0 & 1 \end{bmatrix}$ for the proportional, integral and derivative actions respectively. The controller is first tested on the problem of operating range change. The responses obtained before and after the change are shown in Figures 8.16 and 8.17. Note that due to the lower degree of nonlinearity of the system, it is not so much affected by the change as the neural controller case (refer to Figures 8.11 and 8.17). The PID controller is next tested on the problem of dynamics change. The results obtained are shown in Figures 8.18 and 8.19 and the performance data obtained are also given in Table 8.2. It can be seen that before the change, the control performance of the PID controller is slightly better than that of the neural controller (refer to Figures 8.13 and 8.18, and Table 8.2). In particular, the offset is smaller due to the integral action.

8.4 Discussion

From the simulation results obtained, the following observations can be made:

(i) The proposed neural controller can be trained automatically to give a satisfactory performance in controlling a nonlinear multivariable plant (see Figures 8.10 and 8.13). On the other hand, for the PID controller to produce similar results (see Figures 8.16 and 8.18), the gain matrix have to be manually tuned by the system designer;

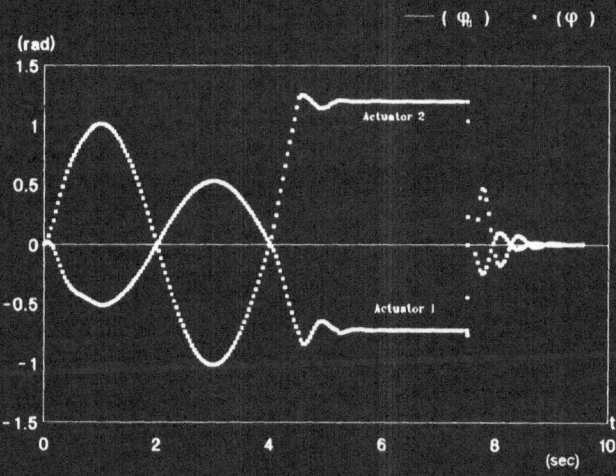

Figure 8.13 Control responses with a well trained controller (stable operating conditions; invariant plant)

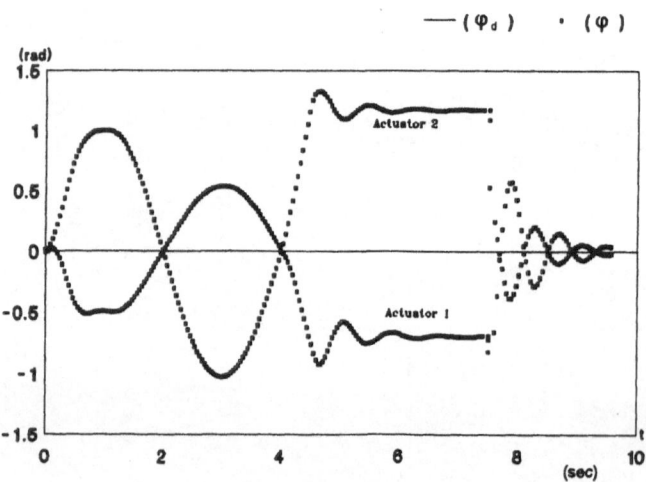

Figure 8.14 Control responses immediately after a dynamics change

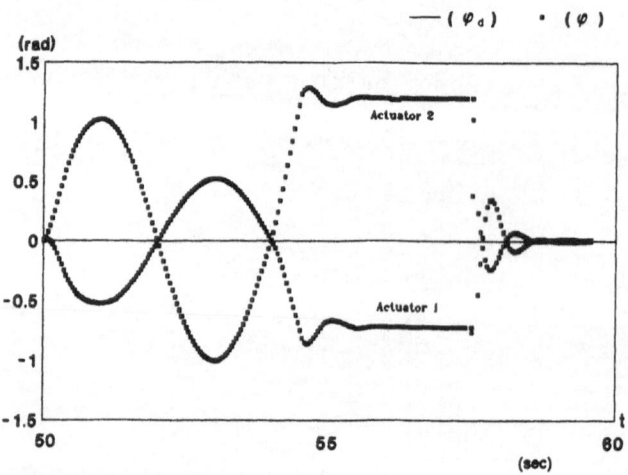

Figure 8.15 Control responses after adaptation to the dynamics change

Table 8.2 Performances of neural controller and PID controller

Performance / Controller			Maximum Overshoot (Ramp) %	Maximum Overshoot (Step) %	5%-Setting Time (Step) sec
Neural Controller	Before dynamics change	(φ_1)	13.5	63.1	1.02
		(φ_2)	2.4	20.1	0.70
	Before adapting to change	(φ_1)	25.7	77.6	1.30
		(φ_2)	8.3	31.3	1.32
	After adapting to change	(φ_1)	17.8	46.7	0.58
		(φ_2)	5.3	19.6	0.54
PID Controller	Before dynamics change	(φ_1)	25.7	47.1	0.66
		(φ_2)	9.7	16.7	0.62
	After dynamics change	(φ_1)	39.1	61.6	1.86
		(φ_2)	16.5	26.4	1.14

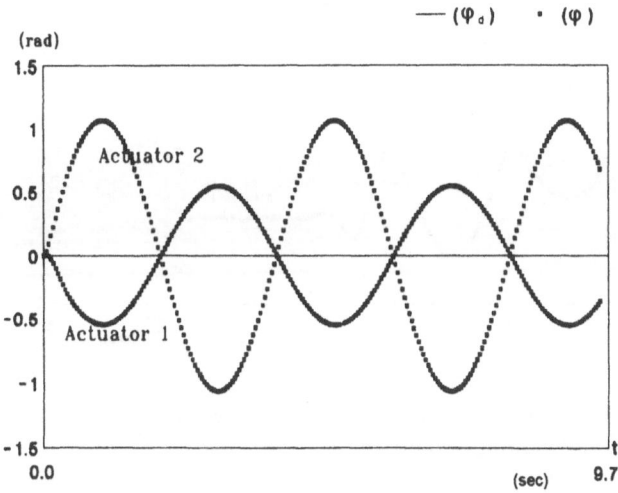

Figure 8.16 Control responses obtained with PID controller (before operating range change)

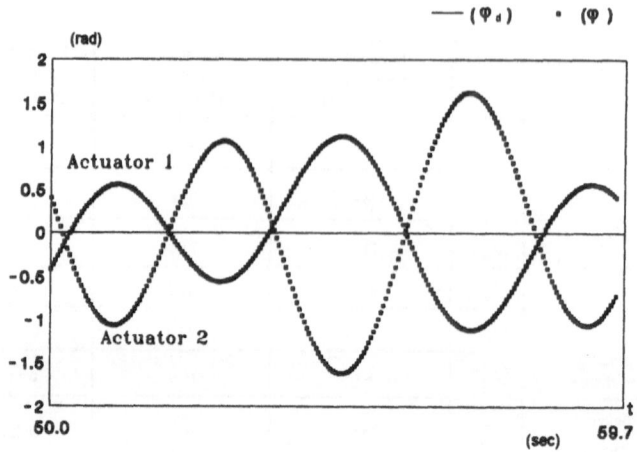

Figure 8.17 Control responses obtained with PID controller (after operating range change)

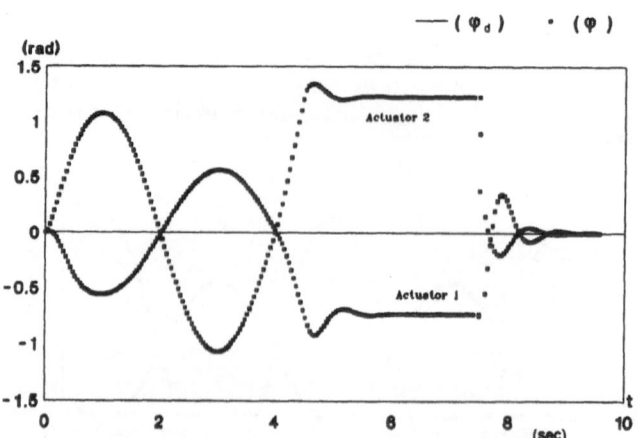

Figure 8.18 Control responses obtained with PID controller (before dynamics change)

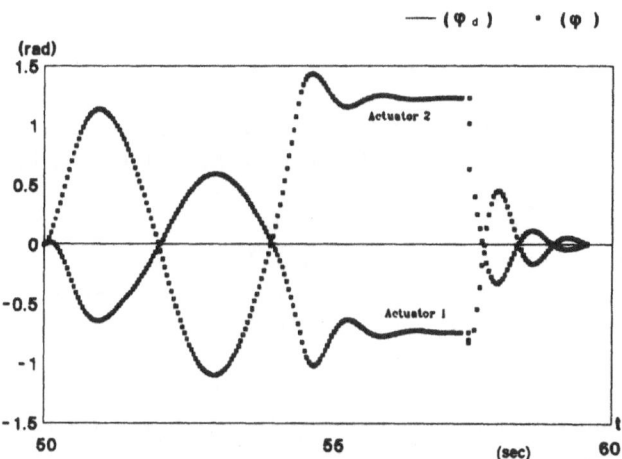

Figure 8.19 Control responses obtained with PID controller (after dynamics change)

(ii)   The neural controller proves more robust to plant parameter change than the PID controller even before the adaptation process is underway (compare Figures 8.14 and 8.19, and also refer to Table 8.2);

(iii)   The adaptation mechanism is effective in improving the performance of the neural controller following a change in the plant parameters (compare Figures 8.14, 8.15 and 8.19, and also refer to Table 8.2).

8.5 Summary

This Chapter has described a new approach to the neural control of a multi-joint robot. The approach can be employed in the control of general multi-input multi-output nonlinear dynamic systems. Simulations have been conducted to investigate the application of the control approach to a two-joint Scara robot.

References

Ciliz, M.K. (1990) *Artificial neural network based control of nonlinear systems with application to robotic manipulators*, PhD thesis, Electrical Engineering, Syracuse University, USA.

Colombano, S.P., Compton, M., and Bualat, M. (1991) Goal directed model inversion: adaptation to unexpected model changes, *Proc. 4th Int. Conf. on Neural Networks and Their Applications (NEURO-NIMES 91)*, Nimes, France, 271-278.

Craig, J.J. (1989) *Introduction to robotics: Mechanics and Control*, Reading, MA: Addison-Wesley.

Craig, J.J., Hsu, P., and Sastry, S.S. (1987) Adaptive control of mechanical manipulators, *Int. J. of Robotics Research*, 6(2), 16-28.

D'Souza, A.F. (1988) *Design of control systems*, Englewood Cliffs, NJ: Prentice-Hall.

Gerald, C.F. and Wheatley, P.D. (1989) *Applied numerical analysis*, Reading, England: Addison-Wesley.

Goldberg, K. and Pearlmutter, B. (1988) Using neural network to learn the dynamics of the CMU Direct-Drive Arm II, Technical Report, CMU-CS-88-160, Carnegie Mellon University.

Kawato, M., Uno, Y., Isobe, M., Suzuki, R. (1988) Hierarchical neural network model for voluntary movement with application to robotics, *IEEE Control Systems Magazine*, April 1988, 8-16.

Khosla, P.K. (1986) *Real-time control and identification of direct-drive manipulators*, PhD thesis, Department of Electrical and Computer Engineering, Carnegie Mellon University.

Kung, S.Y. and Hwang, J.N. (1991) Neural network architectures for robotics applications, *IEEE Trans. on Robotics and Automation*, 5(5), 641-657.

Kuo, B. (1982) *Automatic Control Systems*, Englewood Cliffs, NJ: Prentice-Hall.

Maniko, H., Furuya, N., Soma, K., and Chin, E. (1980) Research and Development of the Scara robot, *Proc. 4th Int. Conf. on Production Engineering*, Tokyo.

Miller, III, W.T., Glanz, F.H., and Klaft, III, L.G. (1987) Application of a general learning algorithm to the control of robotic manipulators, *Int. J. of Robotics Research*, 6(2), 84-98.

Miller, W.T., Sutton, R.S., and Werbos, P.J. (1990) *Neural Networks for Control*, Cambridge, MA: MIT Press.

Oh, S. J. (1993) *Identification and Control of Dynamic Systems Using Neural Networks*, PhD thesis, School of Electrical, electronic and Systems Engineering, University of Wales Cardiff, UK.

Ozaki, T., Suzuki, T., Furuhashi, T., Okuma, S., and Uchikawa, Y. (1991) Trajectory control of robotic manipulators using neural networks, *IEEE Trans. on Industrial Electronics*, 38(3), 641-657.

Pham, D.T. and Oh, S.J. (1992a) A recurrent backpropagation neural network for dynamic system identification, *Journal of Systems Engineering*, 2(4), 213-223.

Pham, D.T. and Oh, S.J. (1992b) Identification of plant inverse dynamics using neural networks, Research Report, Dept. of ELSYM, University of Wales.

Pham, D.T. and Oh, S.J. (1993) Adaptive control of dynamic systems using neural networks, *Proc. IEEE-SMC Conf., Systems Engineering in the Service of Human,* Le Touquet, France, Oct. 1993, 4, 97-102.

Psaltis, D., Sideris, A., and Yamamura, A. (1989) A multilayered neural network controller, *IEEE Control Systems Magazine*, April 1989, 17-21.

Slotine, J.J. and Li, W. (1987) On the adaptive control of robot manipulators, *Int. J. of Robotics Research*, 6(3), 49-59.

Spong, M.W. and Vidyasagar, M. (1989) *Robot Dynamics and Control*, NY: John Wiley & Sons.

Tomizuka, M., Horowitz, R., Anwar, G., and Jia, Y.L. (1988) Implementation of adaptive techniques for motion control of robotic manipulators, *ASME J. Dynamical Systems, Measurement and Control*, 110, 62-69.

Appendix A Introduction to Some Conventional Techniques of Identification, Prediction and Control

This appendix summarises the main conventional techniques for identification, time-series prediction, and control. The material provides a background against which the neural-network-based techniques covered in the book may be contrasted.

A1 Identification

A1.1 Concepts

The knowledge of the model of a system is useful in studying its behaviour. The model can be mental, graphic, or mathematical. Mathematical models are the most useful in this respect. To build a mathematical model of a system, one can use the physical laws that govern the system's behaviour. Alternatively, one can observe the signals produced by the system to known inputs and find a model that best reproduces the observed data. The former approach is called modelling; the latter is called identification. Identification is necessary when there is not sufficient information about the system for it to be accurately modelled.

Once a model has been obtained, it can be used for analysis of a system's properties, for prediction, and for controller design.

A1.2 Identification Approaches

Existing identification approaches can be classified into two groups: nonparametric and parametric approaches.

Nonparametric approaches: These approaches include transient analysis, frequency analysis, correlation analysis, and spectral analysis.   The transient analysis approach is easy to apply. It involves applying step or impulse inputs to a system and produces a step response or an impulse response as a model. However, transient analysis is sensitive to noise and can only give a very approximate model. The frequency analysis approach is based on the use of sinusoids as inputs to a system. It requires rather long identification experiments, especially if correlation is employed to reduce sensitivity to noise. The resulting model is a frequency response. It can be presented as a Bode plot or an equivalent transfer function. Correlation analysis is generally based on using white noise as input. It gives a weighting function as the resulting model. It is insensitive to additive noise on the output signal. Spectrum analysis can be applied with rather arbitrary inputs.   The model is obtained as a Bode plot (or another equivalent form).

Nonparametric methods are easy to apply but give only moderately accurate models. If high accuracy is required a parametric method has to be used. In such cases, nonparametric methods can be used to obtain a first approximate model, which may give useful information on how to apply the parametric method.

Parametric approaches: There are several parametric approaches. A simple example of a parametric approach is linear regression using least squares.

When applying linear regression, the following model is considered:

$$A(z^{-1})y(k) = B(z^{-1})u(k) + \varepsilon(k) \qquad\qquad (A.1)$$

where

$$A(z^{-1}) = 1 + a_1 z^{-1} + \ldots + a_l z^{-l} \quad \text{and} \quad B(z^{-1}) = b_1 z^{-1} + \ldots + b_l z^{-l}.$$

The above equations can be rewritten as

$$y(k) = \varphi^T(k)\theta + e(k) \tag{A.2}$$

where

$$\varphi^T(k) = (-y(k-1)\ldots -y(k-l) \quad u(k-1)\ldots u(k-m))$$
$$\theta = (a_1 \ldots a_l \quad b_1 \ldots b_m)^T$$

The parameter vector which minimises the sum of squared errors

$$V_N(\theta) = \frac{1}{N}\sum_{k=1}^{N} e^2(k) \tag{A.3}$$

is given by

$$\hat{\theta} = \left[\frac{1}{N}\sum_{k=1}^{N}\varphi(k)\varphi^T(k)\right]^{-1}\left[\frac{1}{N}\sum_{k=1}^{N}\varphi(k)y(k)\right] \tag{A.4}$$

This is the least-square (LS) method.

The estimate $\hat{\theta}$ is consistent ($\hat{\theta}$ tends to true parameter vector θ_0 as N tends to infinity) if

(i) $E\varphi(k)\varphi^T(k)$ is non-singular and
(ii) $E\varphi(k)e(k) = 0$

Condition (i) is satisfied in most cases, but condition (ii) is not.

The LS method is simple to use, but gives consistent parameter estimates only under rather restricted conditions. In some cases, the lack of accuracy may be tolerable. If the signal-to-noise ratio is large, the bias in parameter estimates will be small. However, in some cases, even small biases are not allowed. To improve the LS method, the prediction error method [Soderstrom and Stoica, 1989] can be used.

A2 Time Series Prediction

A2.1 Concepts

Time-series prediction involves using the available observations from a time series at time k to forecast its value at some future time $k+l$.

As with system identification, the time-series prediction approach based on mathematical models is well established. Deterministic models can be found if parameters are known exactly. It is not possible to obtain deterministic models if there are too many uncertain factors in a process. In such cases, stochastic models are used to yield the probability of a future value lying between two specific limits.

Stationary models are an important class of stochastic models for describing time series. They are suitable for characterising processes which remain in equilibrium about a constant mean level. However, in industry and business, non-stationary models which assume no constant means are often required.

Autoregressive model: In this stationary model, a finite linear aggregate of previous values of the process and an impulse u_k are used to represent the current value of the process.

Let the values of a process at time instants k, k-1, k-2, ... be represented by y_k, y_{k-1}, y_{k-2}, ... Also let $\bar{y}_k$, $\bar{y}_{k-1}$, $\bar{y}_{k-2}$, ... be the deviations from the mean value μ, $\bar{y}_k = y_k - \mu$. Then the autoregressive (AR) process (of order p) is described by

$$\Phi(B)\bar{y}_k = u_k \tag{A.5}$$

where $\Phi(B) = 1 - \Phi_1 B - \Phi_2 B^2 - ... - \Phi_p B^p$ and $B\bar{y}_k = \bar{y}_{k-1}$.

Moving average model: The moving average (MA) model is another important stationary model employed in practice. In a moving average model of order q, a finite linear aggregate of previous values of impulse u_k is employed to represent the current value of the process, viz:

$$\bar{y}_k = u_k - \theta_1 u_{k-1} - \theta_2 u_{k-2} . \quad . \quad . - \theta_q u_{k-q} \tag{A.6}$$

or

$$\bar{y}_k = \theta(B)u_k \tag{A.7}$$

where

$$\bar{y}_k = y_k - \mu, \quad \mu: \text{mean value}$$

and

$$\theta(B) = 1 - \theta_1 B - \theta_2 B^2 \ldots \ldots - \theta_q B^q \tag{A.8}$$

Autoregressive moving average (ARMA) model: This more general stationary model includes both the autoregressive part and moving average part, which leads to

$$\Phi(B)\bar{y}_k = \theta(B)u_k \tag{A.9}$$

where $\Phi(B)\bar{y}_k$ and $\theta(B)u_k$ are given by (A.5) and (A.7) respectively.

Non-stationary models: Many time series encountered in practice are non-stationary and typically do not vary about a fixed mean. To handle such series, a new autoregressive operator is introduced, viz:

$$\varphi(B) = \Phi(B)(1 - B)^d \tag{A.10}$$

where $\Phi(B)$ is a stationary operator. The aim of introducing such an operator is to transform the series into a stationary one by taking the dth difference of the process. The new model is

$$\varphi(B)y_k = \Phi(B)(1 - B)^d y_k = \theta(B)u_k \tag{A.11}$$

The above model can be employed for describing stationary and non-stationary times series. It is called the autoregressive integrated moving average (ARIMA) model of order (p, d, q) where p, d, and q are the same as those used in Eqs (A.5), (A.6) and (A.11).

A2.2 Model Building

Three stages are used iteratively to build a model.

Model identification: This is performed based on observations of the time series. In general, this involves selecting a suitable subclass of models from the general ARIMA family.

Model estimation: This is carried out to find the parameters of the selected models. The methods adopted are usually Likelihood or Bayesian methods.

Model diagnosis: This is undertaken to find out inadequacies of the model and improve it. In general, this is achieved by analysing the autocorrelation function of the residuals of the parameters.

A2.3 Prediction

The aim is to forecast the value y_{k+l} ($l \geq 1$) of the series.

There are three explicit ways to express an observation y_{k+l}. The first is directly based on the difference equation:

$$y_{k+l} = \varphi_1 y_{k+l-1} + \ldots + \varphi_{p+d} y_{k+l-p-d} - \theta_1 u_{k+l-1} - \ldots - \theta_q u_{k+l-q} + u_{k+l}$$

$$(A.12)$$

The second method of expression uses an infinite weighted sum of current and previous impulses u_j:

$$y_{k+l} = \sum_{j=-\infty}^{k+l} \psi_{k+l-j} u_j = \sum_{j=0}^{\infty} \psi_j u_{k+l-j} \qquad (A.13)$$

where $\psi_0 = 1$. The ψ weights may be obtained by solving the following equation:

$$\varphi(B)(1 + \psi_1 B + \psi_2 B^2 + \ldots) = \theta(B) \qquad (A.14)$$

The third method is based on an infinite weighted sum of previous observations, plus a random impulse:

$$y_{k+l} = \sum_{j=1}^{\infty} \pi_j y_{k+l-j} + u_{k+l} \tag{A.15}$$

The π weights can be obtained from equation

$$\varphi(B) = \left(1 - \pi_1 B - \pi_2 B^2 - \ldots\right)\theta(B) \tag{A.16}$$

Derivation of the minimum mean square error prediction: Assume that the aim at time k is to predict $\bar{y}_k(l)$ of y_{k+l} which can be represented as a linear function of current and previous observations y_k, y_{k-1}, y_{k-2}, ... or a linear function of current and previous impulses u_k, u_{k-1}, u_{k-2}, ...

   Suppose the best prediction is

$$\bar{y}_k(l) = \psi_l^* u_k + \psi_{l+1}^* u_{k-1} + \psi_{l+2}^* u_{k-2} + \ldots \tag{A.17}$$

where the weights are to be determined. The prediction error is

$$E[y_{k+l} - \bar{y}_k(l)]^2 = (1 + \psi_1^2 + \ldots + \psi_{l-1}^2)\sigma_u^2 + \sum_{j=0}^{\infty}\{\psi_{l+j} - \psi_{l+j}^*\}^2 \sigma_u^2 \tag{A.18}$$

where σ_u is the standard deviation of u. The prediction error can be minimised by setting $\psi_{l+j}^* = \psi_{l+j}$. Then

$$\begin{aligned} y_{k+l} &= (u_{k+l} + \psi_1 u_{k+l-1} + \ldots + \psi_{l-1} u_{k+1}) + (\psi_l u_k + \psi_{l+1} u_{k-1} + \ldots) \\ &= e_k(l) + \bar{y}_k(l) \end{aligned} \tag{A.19}$$

where $e_k(l)$ is the error of the forecast $\bar{y}_k(l)$ at lead time l .

A3 Dynamic System Control

A3.1 Concepts

An automatic controller can be used in an open-loop system or a closed-loop system. Figure A.1 shows the structure of an open-loop and a closed-loop control system. Both types of systems can be found in practice. Unlike open loop systems, closed loop systems can reject disturbances and provide more accurate control. On the other hand, they can also be affected by stability problems not present in open loop systems.

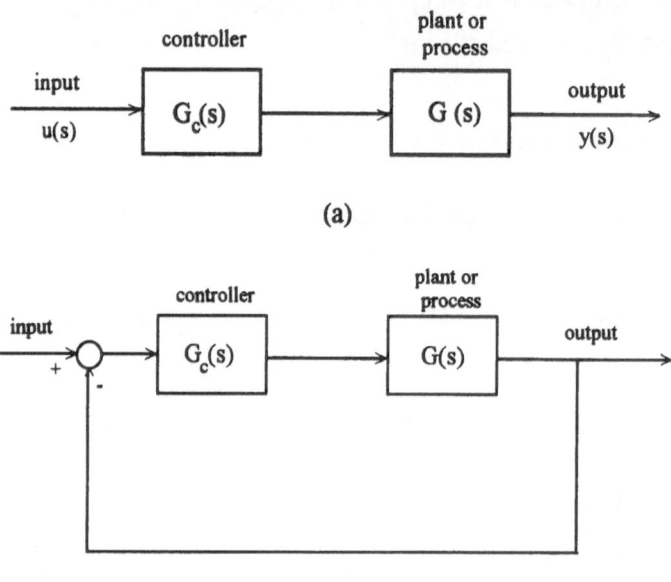

(a)

(b)

Figure A1 Open-loop and closed-loop control systems

A3.2 Control Approaches

Conventional Feedback Control: Conventional feedback control techniques are based on frequency or time domain analysis methods. Processes and controllers are described by transfer functions. When a system $G(s)$ is linear and its parameters are known, conventional feedback controller design techniques can be used to find a suitable controller $G_c(s)$ having a desired output-input relation (closed-loop transfer function).

The closed-loop control performance can be defined by parameters such as the steady-state accuracy, phase margin, and bandwidth. The purpose of controller design is to achieve zero steady-state error, fast response (which is related to the bandwidth), as well as maintain stability (which is related to the phase margin).

A simple but popular controller is the PID controller. By using a fixed controller configuration (proportional part, derivative part, and integral part), the design task reduces to adjusting only three parameters based on simple experiments on the controlled process (Ziegler-Nichols tuning).

Adaptive control approaches: Changes such as shifting the operating point in nonlinear control systems often lead to parameter variations. These variations can affect the system accuracy and stability severely. Through the use of adaptive control, control engineers aim to make the system automatically redesign the controller when parameter changes occur.

There are two important adaptive control schemes: model reference adaptive control (MRAC) and self-tuning regulator. The block diagram of a typical MRAC system is shown in Figure A.2.

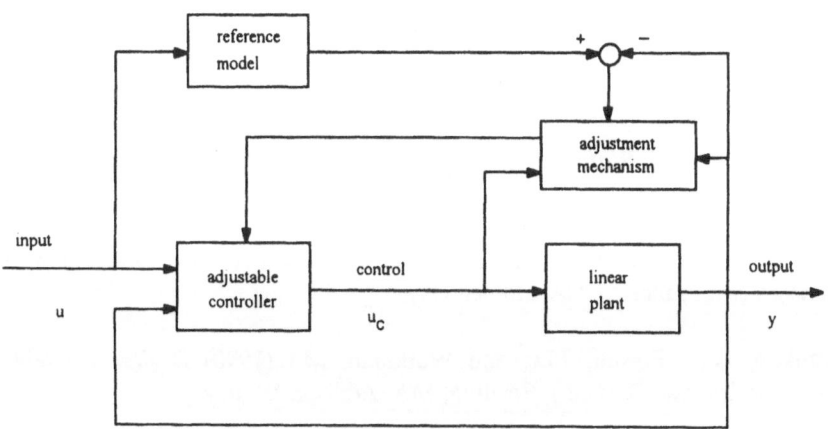

Figure A2 Model reference adaptive control system

A direct MRAC system modifies the controller parameters directly. An indirect MRAC system updates a reference model, and the parameters of that model are then used to compute the controller parameters.

A self-tuning regulator identifies the plant parameters when the system is running. The identified parameters are then employed to redesign the controller. This scheme is called the explicit self-tuning regulator, as shown in Figure A3. The controller parameters can also be identified directly in which case the controller is called a implicit self-tuning regulator.

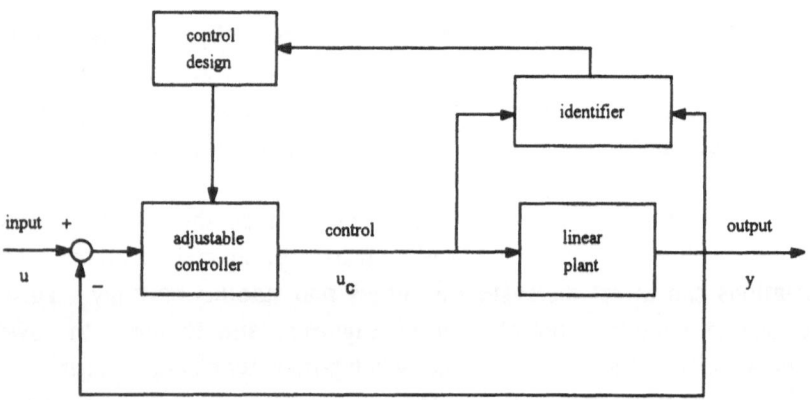

Figure A3   Explicit self-tuning controller

References

Astrom, K.J. and Wittenmark, B. (1989) *Adaptive Control* , Reading, MA: Addison-Wesley.

Box, G.E.P. and Jenkins, G.M. (1978) *Time Series Analysis, Forecasting, and Control*, San Francisco, CA: Holden Day.

Franklin, G.F., Powell, J.D., and Workman, M.L.(1990) *Digital Control of Dynamic Systems* (2nd ed.), Reading, MA: Addison-Wesley.

Ljung, L. (1987) *System Identification: Theory for the User*. Englewood Cliffs, NJ: Prentice-Hall.

Ljung, L. and Soderstrom, T. (1983) *Theory and Practice of Recursive Identification*, Cambridge, MA: MIT Press.

Soderstrom, T. and Stoica, P. (1989) *System Identification*, London: Prentice Hall.

Appendix B Fuzzy Sets and Fuzzy Logic Control

Fuzzy logic control is based on fuzzy set theory [Zadeh, 1965; Zimmermann, 1991]. In the first part of this appendix, fuzzy set theory is introduced and some basic fuzzy set operations presented. Approximate reasoning using fuzzy relations is also explained [Zadeh,1973]. In the second part of this appendix, the basic notions fuzzy logic control are briefly described. The material used is adopted from [Karaboga, 1994].

B1. Fuzzy Set Theory

Fuzzy set theory may be considered an extension of classical set theory. While classical set theory is about "crisp" sets with sharp boundaries, fuzzy set theory is concerned with "fuzzy" sets whose boundaries are "grey".

In classical set theory, an element (u) can either belong or not belong to a set A, i.e. the degree to which element u belongs to set A is either 1 or 0. However, in fuzzy set theory, the degree of belonging of an element u to a fuzzy set A is a real number between 0 and 1. This is denoted by $\mu_A(u)$, the grade of membership of u in A. Fuzzy set A is a fuzzy set in U, the "universe of discourse" or "universe" which includes all objects to be discussed. $\mu_A(u)$ is 1 when u is definitely a member of A and $\mu_A(u)$ is 0 when u is definitely not a member of A. Figures B1(a) and (b) show a crisp set A and fuzzy set A of numbers near to 2.

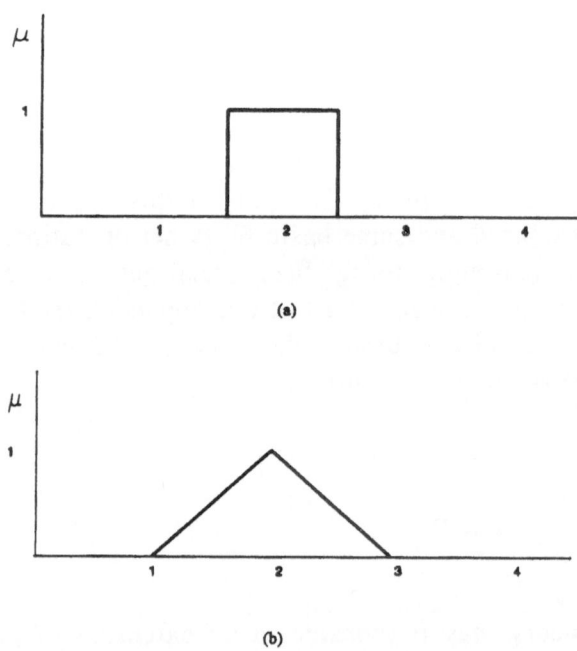

(a)

(b)

Figure B1 (a) Crisp set of numbers "near" to 2; (b) Fuzzy set of numbers "near" to 2.

B1.1 Definitions

Definition 1. A fuzzy set A in U is a set of ordered pairs:

$$A = \{ (u, \mu_A(u)) \mid u \in U \} \tag{B1}$$

Definition 2. The support of a fuzzy set A, $S(A)$, is the crisp set of all $u \in U$ such that $\mu_A(u) > 0$.

Definition 3. The (crisp) set of elements that belong to fuzzy set A at least to a given degree α is called the α-level set.

$$A_{\alpha} = \{ u \in U \mid \mu_A(u) \geq \alpha \} \tag{B2}$$

$A_{\alpha} = \{ u \in U \mid \mu_A(u) > \alpha \}$ is called a "strong α-level set" or "strong α cut".

Definition 4. A fuzzy set A defined in U is called a fuzzy singleton if the support of A contains only one element and its membership value is 1.

Definition 5. If A and B are fuzzy sets in universes U and V, respectively, the Cartesian product of A and B is a fuzzy set in the product space UxV with membership function

$$\mu_{A \times B}(u, v) = Min \{ \mu_A(u), \mu_B(v) \} \tag{B3}$$

where $(u, v) \in (U \times V)$

B1.2 Basic Operations on Fuzzy Sets

Fuzzy sets are defined by membership functions. Therefore, set operations are usually defined in terms of membership functions. The most commonly adopted definitions are those used by Zadeh [Zadeh,1965]. They are given below for the intersection, union and complement operations.

Definition 6. The intersection C of fuzzy sets A and B is defined by

$$\mu_C(u) = Min \{ \mu_A(u), \mu_B(u) \}, \qquad u \in U \tag{B4}$$

Definition 7. The union D of fuzzy sets A and B is given by

$$\mu_D = Max \{ \mu_A(u), \mu_B(u) \}, \qquad u \in U \qquad \text{(B5)}$$

Definition 8. The complement $\bar{A}$ of fuzzy set A is defined by

$$\mu_{\bar{A}}(u) = 1 - \mu_A(u), \qquad u \in U \qquad \text{(B6)}$$

Let A and B be the fuzzy sets of numbers near to 2 and near to 3 (Figure B2(a)). The intersection and the union of these two sets are given in Figures B2(b) and (c), respectively. Figure B2(d) shows the complement of fuzzy set A.

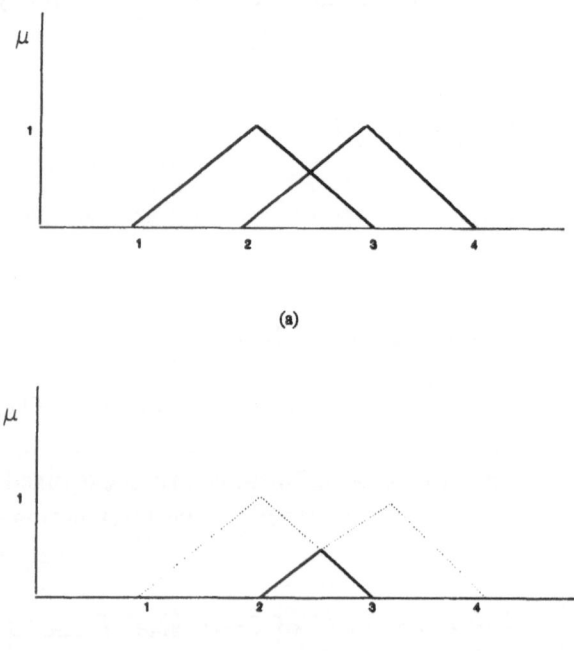

(a)

(b)

Figure B2 (a) Fuzzy sets representing numbers "near" to 2 and 3; **(b)** Intersection of the two fuzzy sets in **(a)**.

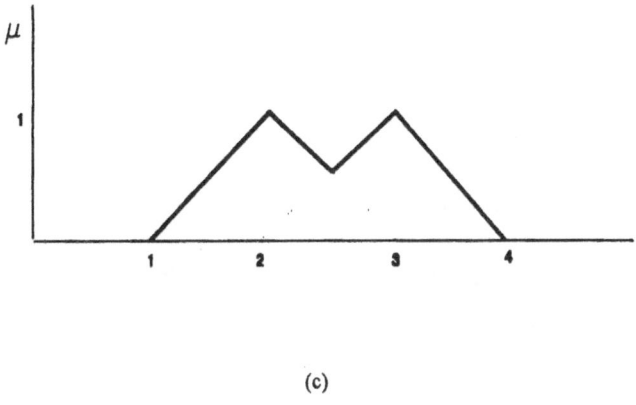

(c)

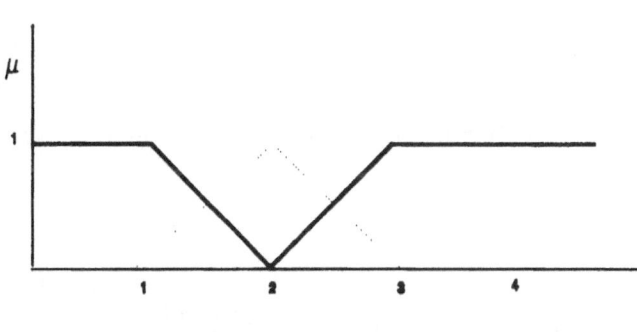

(d)

Figure B2 (c) Union of the fuzzy sets in Figure B2(a); **(d)** Complement of the fuzzy set representing numbers "near" to 2.

B1.3 Fuzzy Relations

A fuzzy system consists of a set of rules called "fuzzy rules". A simple fuzzy rule is of the following form:

"IF $\underset{\sim}{A}_i$ THEN $\underset{\sim}{B}_i$"

where the antecedent $\underset{\sim}{A}_i$ and the consequent $\underset{\sim}{B}_i$ are qualitative expressions.

An example of a fuzzy rule for a fuzzy logic controller is

"IF the error between the desired and actual outputs of the process is Positive_Big
THEN the change in the controller output (the input of the process) is Negative_Big"

Here, A_i represents "error is Positive_Big" and B_i "change in controller output is Negative_Big".   Such a rule expresses a conditional relation R_i between A_i and B_i . R_i is generally identified with the previously defined Cartesian product, i.e.

$$R_i = A_i \times B_i = \sum_{UxV} \mu_{R_i} (u, v) / (u, v) \tag{B7}$$

R_i can be expressed as a two dimensional matrix with elements $\mu_{R_i} (u,v)$:

$$\mu_{R_i} (u,v) = \begin{bmatrix} \mu_{R_i} (u_1,v_1) & \cdots & \mu_{R_i} (u_1,v_n) \\ \vdots & & \vdots \\ \mu_{R_i} (u_n,v_1) & \cdots & \mu_{R_i} (u_n,v_n) \end{bmatrix} \tag{B8}$$

If "n" fuzzy rules can be expressed as relations R_1 to R_n, an overall fuzzy relation R representing all rules can be constructed by combining these individual relations.   By employing the union operation, R can be obtained thus :

$$R = \bigcup_{i=1}^{n} R_i \tag{B9}$$

B1.4 Compositional Rule of Inference

This rule enables a fuzzy system to infer an output corresponding to an arbitrary input not explicitly covered by the individual fuzzy rules using the separate relation matrices R_1 to R_n or the overall relation matrix R. This rule therefore describes a method of approximate reasoning [Zadeh,1975].

In its simplest form, for one input variable and one output variable, the rule is expressed by the following equation:

$$b = a \circ R \tag{B10}$$

Here, a is the arbitrary input and is not the antecedent of any of the relations R_1 to R_n. b is the inferred process output given as:

$$b = \sup (a * R) \tag{B11}$$

The compositional rule of inference can be regarded as a generalised "modus ponens" rule [Lee,1990a]. This rule permits one to infer that the output of a process is b', based on the relation "**If** x *is* a, **Then** y *is* b" and the fact "x *is* a'". In the case of the compositional rule of inference, the variables are fuzzy and in the case of the conventional modus ponens rule the variables are crisp.

The compositional rule of inference has two advantages. First, it enables approximate and uncertain inputs to be dealt with. Second, it allows the number of relations required to describe the behaviour of a process to be kept small because it is not necessary to cover all possible inputs.

B2 Fuzzy Logic Controllers

B2.1 Basic Structure of a Fuzzy Logic Controller

Fuzzy set theory was first applied to the design of a controller for a dynamic plant by Mamdani in 1974 [Mamdani,1974]. Controllers which employ fuzzy set concepts are called fuzzy logic controllers

(FLCs). Figure B3 shows the basic structure of an FLC. It consists of four principal units. These are the fuzzification, knowledge base, decision-making (computation) and defuzzification units. Since data manipulation in an FLC is based on fuzzy set theory, a fuzzification process is required to convert the measured "crisp" inputs to "fuzzy" values. The fuzzification unit first maps the measured values of input variables into corresponding universes of discourse. It then converts the mapped input data into fuzzy sets based upon fuzzy values, such as Positive_Big (PB), Negative_Small (NS) etc.

The knowledge base contains a set of rules or a relation matrix representing those rules and the information regarding the normalisation and/or discretization of universes of discourse, fuzzy partitions of input and output spaces, and membership functions defining fuzzy values. Definitions used for the manipulation of fuzzy data are also stored in this unit.

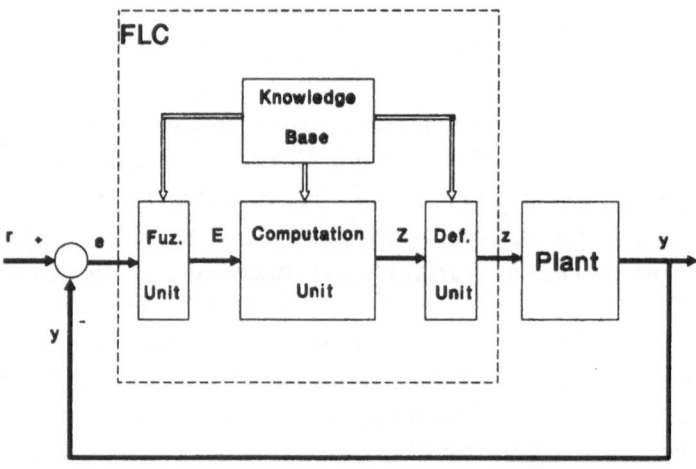

Figure B3 Basic structure of an FLC using only input-output data

The relation between the input and output of a simple fuzzy logic controller is represented as a set of fuzzy rules or the corresponding overall fuzzy relation matrix. In other words, those rules/relation matrices form the control strategy for the process. The main problem in FLC design is how to obtain those fuzzy rules.

The decision-making unit simulates the inference mechanism in humans. It produces fuzzy control actions using fuzzy implication which is the expression of a fuzzy relation and the rule of inference in fuzzy logic. Researchers have experimented with different methods for implication and inference to determine the most suitable methods for fuzzy control problems [Lee,1990b]. For example, using eight intuitive criteria, Lee tested seven widely employed implication and inference methods and concluded that the following two are the best for FLCs.

1. Max-min Rule. This rule is also called Mamdani's minimum operation rule since it was first applied by Mamdani [Mamdani,1974]. Using this rule, for the implication and inference, the following can be written :

$$\mu_{\underset{\sim}{R}}(u,v) = Min\ (\mu_{\underset{\sim}{A}}(u),\ \mu_{\underset{\sim}{B}}(v)) \tag{B12}$$

$$\mu_{\underset{\sim}{B'}}(u,v) = \underset{U}{Max}\ \{\ Min\ (\mu_{\underset{\sim}{A'}}(u),\ \mu_{\underset{\sim}{R}}(u,v))\ \} \tag{B13}$$

2. Max-product Rule. This rule is obtained by replacing the Min operation in the previous rule with the product operation. It was proposed by Larsen [Larsen,1980].

$$\mu_{\underset{\sim}{R}}(u,v) = (\mu_{\underset{\sim}{A}}(u) \cdot \mu_{\underset{\sim}{B}}(v)) \tag{B14}$$

$$\mu_{\underset{\sim}{B'}}(u,v) = \underset{U}{Max}\ \{\ (\mu_{\underset{\sim}{A'}}(u) \cdot \mu_{\underset{\sim}{R}}(u,v))\ \} \tag{B15}$$

As before, in the above four equations, $u \in U$, $v \in V$ and $(u,\ v) \in U \times V$, A and B are the fuzzy values of the input and output variables in a control rule. A' is any fuzzy input to the controller and B' is the corresponding fuzzy output.

The output of the decision-making unit is a fuzzy set. However, a deterministic value is generally required as the input to the process. That is, an interface unit between the process and the decision-making unit is necessary. Several procedures have been proposed for the defuzzification task. The following two are commonly used in control applications.

1. Average of maxima criterion.   The deterministic output z is calculated by taking the weighted average of elements v_i having the locally highest membership values, $\mu_H(v_i)$ viz.

$$z = (\sum_{i=1}^{n} \mu_H(v_i) \, (v_i)) \, / \, (\sum_{i=1}^{n} \mu_H(v_i)) \qquad\qquad \text{(B16)}$$

where v_i is the support value at which the membership function reaches the local maximum value μ_H, and n is the number of such support values.

2. Centre of area method.   This is a popular defuzzification method.   It generally gives a better steady state performance than with other methods [Yamazaki,1982].   The crisp output is obtained using the following formula:

$$z = (\sum_{i=1}^{n} \mu(v_i) \, (v_i)) \, / \, (\sum_{i=1}^{n} \mu(v_i)) \qquad\qquad \text{(B17)}$$

where n is the number of support values of the fuzzy set, v_i is a support value and μ is the membership degree of v_i.

References

Karaboga, D. (1994) *Design of Fuzzy Logic Controllers Using Genetic Algorithms*, PhD thesis, University of Wales, Cardiff.

Larsen, P.M. (1980) Industrial application of fuzzy logic control, *J. Man Mach. Studies*, 12(1), 3-10.

Lee, C.C. (1990a) Fuzzy logic in control systems: Fuzzy logic controller, Part I, *IEEE Trans. on Systems, Man and Cybernetics*, 20(2), 404-418.

Lee, C.C. (1990b) Fuzzy logic in control systems: Fuzzy logic controller, Part II, *IEEE Trans. on Systems, Man, and Cybernetics*, 20(2), 419-435.

Mamdani, E.H. (1974) Applications of fuzzy algorithms for control of simple dynamic plant, *Proc. IEE*, **121**(12), 1585-1588.

Yamazaki, T. (1982) *An Improved Algorithm for a Self Organising Controller*, Ph.D. Thesis, Queen Mary College, University of London, U.K.

Zadeh, L.A. (1965) Fuzzy sets, *Inform. Contr.*, **8**, 338-353.

Zadeh, L.A. (1973) Outline of a new approach to the analysis complex systems and decision process, *IEEE Trans. on Systems, Man and Cybernetics*, **3**, 28-44.

Zadeh, L. A. (1975) The concept of a linguistic variable and its application to approximate reasoning-III, *Information Sciences*, **9**, 43-80.

Zimmermann, H.-J. (1991) *Fuzzy Set Theory and Its Applications*, 2nd ed., Dardrecht-Boston: Kluwer.

Appendix C Genetic Algorithms

This appendix provides a brief introduction to genetic algorithms.

C1 Background

Conventional search techniques, such as hill-climbing, are often incapable of optimising non-linear or multi modal functions. In such cases, a random search method is generally required. However, undirected search techniques are extremely inefficient for large domains. A genetic algorithm (GA) is a directed random search technique, invented by Holland [Holland,1975], which can find the global optimal solution in complex multi-dimensional search spaces. A GA is modelled on natural evolution in that the operators it employs are inspired by the natural evolution process. These operators, known as genetic operators, manipulate individuals in a population over several generations to improve their fitness gradually. As discussed in the next section, individuals in a population are likened to chromosomes and usually represented as strings of binary numbers.

The evolution of a population is described by the "schema theorem" [Holland,1975]. A schema represents a set of individuals, i.e. a subset of the population, in terms of the similarity of bits at certain positions of those individuals. For example, the schema 1*0* describes the set of individuals whose first and third bits are 1 and 0, respectively. Here, the symbol * means any value would be acceptable. In other words, the values of bits at positions marked * could be either 0 or 1. A schema is characterised by two parameters: defining length and order. The defining length is the length between the first and last bits with fixed values. The order of a schema is the number of bits with specified values. According to the schema theorem, the distribution of a schema through the population from one generation to the next depends on its order, defining length and fitness.

GAs do not use much knowledge about the problem to be optimised and do not deal directly with the parameters of the problem. They work with codes which represent the parameters. Thus, the first issue in a GA application is how to code the problem under study, i.e. how to represent the problem parameters. GAs operate with a population of possible solutions, not only one possible solution. The second issue is the creation of a set of possible solutions at the start as the initial population. The third issue in a GA application is how to select or devise a suitable set of genetic operators. Finally, as with other search algorithms, GAs have to know the quality of already found solutions to improve them further. Therefore, there is a need for an interface between the problem environment and the GA itself. The design of this interface can be regarded as the fourth issue.

C2 Representation

The parameters to be optimised are usually represented in a string form since genetic operators are suitable for this type of representation. The method of representation has a major impact on the performance of the GA. Different representation schemes might cause different performances in terms of accuracy and computation time.

There are two common representation methods for numerical optimisation problems [Michalewicz,1992]. The preferred method is the binary string representation method. The reason for this method being popular is that the binary alphabet offers the maximum number of schemata per bit compared to other coding techniques. Various binary coding schemes can be found in the literature, for example, Uniform coding, Gray scale coding, etc. The second representation method is to use a vector of integers or real numbers with each integer or real number representing a single parameter.

When a binary representation scheme is employed, an important step is to decide the number of bits to encode the parameters to be optimised. Each parameter should be encoded with the optimal number of bits covering all possible solutions in the solution space. When too few or too many bits are used the performance can be adversely affected.

C3 Creation of Initial Population

At the start of optimisation, a GA requires a group of initial solutions. There are two ways of forming this initial population. The first consists of using randomly produced solutions created by a random number generator, for example. This method is preferred for problems about which no a priori knowledge exists or for assessing the performance of an algorithm.

The second method employs a priori knowledge about the given optimisation problem. Using this knowledge, a set of requirements are obtained and solutions which satisfy those requirements are collected to form an initial population. In this case, the GA starts the optimisation with a set of approximately known solutions and therefore convergence to an optimal solution takes less time than with the previous method.

C4 Genetic Operators

The flowchart of a simple GA is given in Figure C1. There are basically four genetic operators, selection, crossover, mutation and inversion. Some of these operators were inspired by nature. In the literature, many versions of these operators can be found. It is not necessary to employ all of these operators in a GA because each operates independently of the others. The choice or design of operators depends on the problem and the representation scheme employed.   For instance, operators designed for binary strings cannot be directly used on strings coded with integers or real numbers.

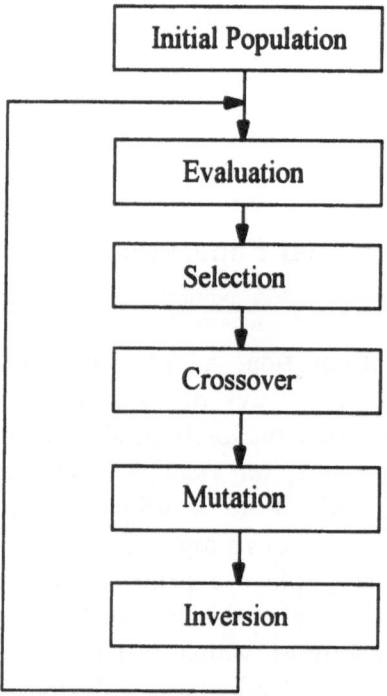

Figure C1 Flowchart of a basic algorithm

C4.1 Selection

The aim of the selection procedure is to reproduce more of individuals whose fitness values are higher than those whose fitness values are low. The selection procedure has a significant influence on driving the search towards a promising area and finding good solutions in a short time. However, the diversity of the population must be maintained to avoid premature convergence and to reach the global optimal solution. In GAs there are mainly two selection procedures: proportional selection, also called stochastic selection, and ranking-based selection [Whitely,1989].

Proportional selection is usually called "Roulette Wheel" selection, since its mechanism is reminiscent of the operation of a Roulette Wheel. Fitness values of individuals represent the widths of slots on the wheel. After a random spinning of the wheel to select an individual for the next generation, slots with large widths representing individuals with high fitness values will have a higher chance to be selected.

One way to control rapid convergence is to control the range of trials allocated to any single individual, so that no individual receives many offspring. The ranking system is one such alternative selection algorithm. In this algorithm, each individual receives an expected number of offspring which is based on the rank of his performance and not on the magnitude [Baker,1985].

C4.2 Crossover

This operation is considered the operation that makes the GA different from other algorithms, such as dynamic programming, etc. It is used to create two new individuals (children) from two existing individuals (parents) picked from the current population by the selection operation. There are several ways of doing this. Some common crossover operations are one-point crossover, two-point crossover, cycle crossover and uniform crossover.

One-point crossover is the simplest crossover operation. Two individuals are randomly selected as parents from the pool of individuals formed by the selection procedure and cut at a randomly selected point. The tails, which are the parts after the cutting point, are swapped and two new individuals (children) are produced. Note that this operation does not change the values of bits. An example of crossover is shown in Figure C2.

Parent 1	1 0 0 | 0 1 0 0 1 1 1 1 0
Parent 2	0 0 1 | 0 1 1 0 0 0 1 1 0
	|
New string 1	1 0 0 | 0 1 1 0 0 0 1 1 0
New string 2	0 0 1 | 0 1 0 0 1 1 1 1 0

Figure C2 Crossover

C4.3 Mutation

In this procedure, all individuals in the population are checked bit by bit and the bit values are randomly reversed according to a specified rate. Unlike crossover, this is a monadic operation. That is, a child string is produced from a single parent string. The mutation operator forces the algorithm to search new areas. Eventually, it helps the GA avoid premature convergence and find the global optimal solution. An example is given in Figure C3.

$$
\begin{array}{ll}
\textbf{Old string} & \textbf{1 1 0 0 | 0 | 1 0 1 1 1 0 1} \\
\textbf{New string} & \textbf{1 1 0 0 | 1 | 1 0 1 1 1 0 1}
\end{array}
$$

Figure C3 Mutation

C4.4 Inversion

This operator is employed for a group of problems, such as the cell placement problem, layout problems and travelling salesman problem. It also operates on one individual at a time. Two points are randomly selected from an individual and the part of the string between those two points is reversed (see Figure C4).

$$
\begin{array}{ll}
\textbf{Old string} & \textbf{1 0 | 1 1 0 0 | 1 1 1 0 1} \\
\textbf{New string} & \textbf{1 0 | 0 0 1 1 | 1 1 1 0 1}
\end{array}
$$

Figure C4 Inversion of a binary string segment

C4.5 Control Parameters

Important control parameters of a simple GA include population size (number of individuals in the population), crossover rate, mutation rate and inversion rate. Several researchers have studied the effect of these parameters on the performance a GA [Schaffer et al.,1989; Grefenstette,1986; Fogarty,1989]. The main conclusions are as follows. A large population size means the simultaneous handling of many solutions and increases the computation time per iteration; however since many samples from the search space are used, the probability of convergence to a global optimal solution is higher than when using a small population size.

The crossover rate determines the frequency of the crossover operation. It is useful at the start of optimisation to discover a promising region. A low crossover frequency decreases the speed of convergence to such an area. If the frequency is too high, it leads to saturation around one solution. The mutation operation is controlled by the mutation rate. A high mutation rate introduces high diversity in the population and might cause instability. On the other hand, it is usually very difficult for a GA to find a global optimal solution with too low a mutation rate.

C4.6 Fitness Evaluation Function

The fitness evaluation unit acts as an interface between the GA and the optimisation problem. The GA assesses solutions for their quality according to the information produced by this unit and not by using directly information about their structure. In engineering design problems, functional requirements are specified to the designer who has to produce a structure which performs the desired functions within predetermined constraints. The quality of a proposed solution is usually calculated depending on how well the solution performs the desired functions and satisfies the given constraints. In the case of a GA, this calculation must be automatic and the problem is how to devise a procedure which computes the quality of solutions.

Fitness evaluation functions might be complex or simple depending on the optimisation problem at hand. Where a mathematical equation cannot be formulated for this task, a rule-based procedure can be constructed for use as a fitness function or in some cases both can be combined. Where some constraints are very important and cannot be violated, the structures or solutions which do so can be eliminated in advance by appropriately designing the representation scheme. Alternatively, they can be given low probabilities by using special penalty functions.

References

Baker, J.E. (1985) Adaptive selection methods for genetic algorithms, *Proc. First Int. Conf. on Genetic Algorithms and Their Applications*, Pittsburgh, PA, 101-111.

Fogarty, T.C. (1989) Varying the probability of mutation in the genetic algorithm, *Proc. Third Int. Conf. on Genetic Algorithms and Their Applications*, George Mason University, 104-109.

Grefenstette, J.J. (1986) Optimization of control parameters for genetic algorithms, *IEEE Trans. on Systems, Man and Cybernetics*, **16** (1), 122-128.

Holland, J.H. (1975) *Adaptation in Natural and Artificial Systems*, Ann Arbor, MI: University of Michigan Press.

Michalewicz, Z. (1992) *Genetic Algorithms + Data Structures = Evolution Programs*, Berlin: Springer-Verlag.

Schaffer, J.D., Caruana, R.A., Eshelman, L.J. and Das, R. (1989) A study of control parameters affecting on-line performance of genetic algorithms for function optimisation, *Proc. Third Int. Conf. on Genetic Algorithms and Their Applications*, George Mason University, 51-61.

Whitely, D. (1989) The GENITOR algorithm and selection pressure: why rank-based allocation of reproductive trials is best, *Proc. Third Int. Conf. on Genetic Algorithms and Their Applications*, George Mason University, 116-123.

Appendix D Program: Feedforward Network for System Identification

```
/********************  MLP.C  ********************

    multilayer perceptron for identification

    backpropagation with momentum, feedforward structure,
    1 input buffer layer, 1 hidden layer (nonlinear), 1 output
            unit (linear).

    program written in Microsoft Quick C (version 1.0)

    ***************************************************/


#include <conio.h>
#include <graph.h>
#include <stdlib.h>
#include <float.h>
#include <math.h>
#include <stdio.h>

#define n0 4                    /* input layer    */
#define n1 6                    /* hidden layer   */
#define n2 1                    /* output layer   */

struct videoconfig vc;

float uk, uk_1, yk1, yk, yk_1;              /* defined variables used */
```

```
float netin0[n0], netin1[n1], netin2[n2];
float bias1[n1], prdlbs1[n1];
float weight1[n1][n0], weight2[n2][n1];
float prdlwt1[n1][n0], prdlwt2[n2][n1];
float actv1[n1], netout;
float lcoef, motum, rms;
float A1, A2, A3, B1, B2, B3;
unsigned seed_int, seed_u;
int epochs;
int eph_pts = 30, rcl_pts = 100;
float inwtsc = 10.0;


float xxo, yyo;
float timegrph, ukgrph, yk1grph, netoutgrph;

main ()
{
  char ch, ch1;
  _setvideomode (_VRES16COLOR);   /* _VRES16COLOR FOR VGA */
  _getvideoconfig (&vc);          /* _ERESCOLOR FOR EGA */
  xxo = 60;
  yyo = vc.numypixels/2.0;
  _setlogorg (xxo,yyo);

  A1 = 1.752821;   A2 = (-0.818731);  A3 = 0.0;
  B1 = 0.011698;   B2 = 0.010942;     B3 = 0.0;

  printf("\n\nlearn ? (y/n)");
  ch = getche();
  if (ch == 'y')
  {
    printf("\n\nlearning coefficient ?");
    scanf("%f", &lcoef);
    printf("\nmomentum ?");
    scanf("%f", &motum);
    printf("\nseed_int ?");
    scanf("%d", &seed_int);
    printf("\nseed_u ?");
    scanf("%d", &seed_u);
    printf("\nepochs of training ?");
```

```
        scanf("%lu", &epochs);

        initialise ();
        learn ();
        savewt ();
    }
    printf ("\nrecall now? (y/n)");
    ch = getche ();
    if (ch == 'y') {
        readwt ();
        recall ();
    }
    printf ("\nexit? (y/n)\n");
    ch1 = getche ();
    if (ch1 = 'y')
    {
        _clearscreen (_GCLEARSCREEN);
        _setvideomode (_DEFAULTMODE);
    }
}

initialise ()                              /* initialise the weights */
{
    int i,j;
    srand (seed_int);
    for (j=0; j<n1; j++)
    {
        bias1[j] = (2.0*(float)rand()/RAND_MAX-1.0)/inwtsc;
        prdlbs1[j] = 0.0;
        for (i=0; i<n0; i++)
        {
            weight1[j][i] = (2.0*(float)rand()/RAND_MAX-1.0)/inwtsc;
            prdlwt1[j][i] = 0.0;
        }
    }
    for (j=0; j<n2; j++)
    {
        for (i=0; i<n1; i++)
        {
            weight2[j][i] = (2.0*(float)rand()/RAND_MAX-1.0)/inwtsc;
```

```
            prdlwt2[j][i] = 0.0;
       }
    }
}

comptout ()                                    /* forward propagation */
{
   register int i, j;
   float ea, eb;
   for (j=0; j<n1; j++)
   {
      netin1[j] = bias1[j];
      for (i=0; i<n0; i++)
           netin1[j] += weight1[j][i] * netin0[i];
      ea = (float)exp((double)netin1[j]);
      eb = (float)exp((double)((-1.0)*netin1[j]));
      actv1[j] = (ea-eb)/(ea+eb);
   }
   for (j=0; j<n2; j++)
   {
      netin2[j] = 0.0;
      for (i=0; i<n1; i++)
           netin2[j] += weight2[j][i] * actv1[i];
      netout = netin2[j];
   }
}

comptwt ()                                     /* update weights */
{
   register int i, j;
   float error1[n1], error2[n2], sum1[n1];
   float delwt1[n1][n0], delwt2[n2][n1];
   float delbs1[n1];
   for (j=0; j<n2; j++)
   {
      error2[j] = yk1 - netout;
      for (i=0; i<n1; i++)
      delwt2[j][i] = lcoef * error2[j] * actv1[i] + motum * prdlwt2[j][i];
   }
   for (j=0; j<n1; j++)
```

```
     {
        sum1[j] = 0.0;
        for (i=0; i<n2; i++)
                sum1[j] += error2[i] * weight2[i][j];
        error1[j] = (1.0+actv1[j])*(1.0-actv1[j])*sum1[j];
        delbs1[j] = lcoef * error1[j] + motum * prdlbs1[j];
        for (i=0; i<n0; i++)
                delwt1[j][i] = lcoef*error1[j]*netin0[i] + motum*prdlwt1[j][i];
     }
     for (j=0; j<n1; j++)
     {
        bias1[j] += delbs1[j];
        prdlbs1[j] = delbs1[j];
        for (i=0; i<n0; i++)
        {
                weight1[j][i] += delwt1[j][i];
                prdlwt1[j][i] = delwt1[j][i];
        }
     }
     for (j=0; j<n2; j++)
     {
        for (i=0; i<n1; i++)
        {
                weight2[j][i] += delwt2[j][i];
                prdlwt2[j][i] = delwt2[j][i];
        }
     }
}

learn ()                                         /* train net */
{
  FILE *fptr;
  register int k;
  int i;
  srand(seed_u);

  for (k=0; k<epochs; k++)
  {
                _clearscreen (_GCLEARSCREEN);
                printf ("%d", k);
```

```
            _setcolor (4);
            _rectangle (_GBORDER, 0,200, eph_pts,-200);

            printf("\nrms = %f", (float)sqrt((double)rms/eph_pts));
            yk = yk_1 = 0.0;  uk_1 = 0.0;
            timegrph = 0.0;  rms = 0.0;
            srand(seed_u);

        for (i=0; i<eph_pts; i++)
        {
            uk = (2.0*(float)rand()/RAND_MAX-1.0)*2.911162;

            netin0[0] = yk;
            netin0[1] = yk_1;
            netin0[2] = uk;
            netin0[3] = uk_1;

            comptout ();

yk1 = A1*yk + A2*yk_1 + A3*yk_1*yk_1 + B1*uk + B2*uk_1 +
B3*uk*uk;

            yk1grph = (-1.0)*yk1*100.0;
            _setcolor (2);
            _setpixel (timegrph, yk1grph);
            netoutgrph = (-1.0)*netout*100.0;
            _setcolor (7);
            _setpixel (timegrph, netoutgrph);

            comptwt ();

            rms += (yk1 - netout)*(yk1 - netout);

            yk_1 = yk; yk = yk1;
            uk_1 = uk;

            timegrph += 1.0;
        }
    }
```

```
}

recall ()                                    /* test net */
{
    FILE *fptr;
    int i, k;
    float basegrph, scalegrph;
    _clearscreen (_GCLEARSCREEN);
    _setcolor (4);
    _rectangle (_GBORDER,0,200,eph_pts,-200);
    srand(seed_u);

    yk = yk_1 = 0.0;  uk_1 = 0.0;
    timegrph = 0.0;  rms = 0.0;

    for (k=0; k<rcl_pts; k++)
    {
        uk = 2.911162*sin(2.0*3.14159*k/15.0);

        netin0[0] = yk;
        netin0[1] = yk_1;
        netin0[2] = uk;
        netin0[3] = uk_1;

        comptout ();

        timegrph = k*(float)eph_pts/rcl_pts;

        yk1 = A1*yk + A2*yk_1 + A3*yk_1*yk_1 + B1*uk + B2*uk_1 +
B3*uk*uk;

        rms += (yk1 - netout)*(yk1 - netout);

        yk1grph = (-1.0)*yk1*100.0;
        _setcolor (2);
        _setpixel (timegrph, yk1grph);
        basegrph = 0.0;
        _setcolor (14);
        _setpixel (timegrph+2.0, basegrph);
        netoutgrph = (-1.0)*netout*100.0;
```

```
        _setcolor (7);
        _setpixel (timegrph, netoutgrph);

        yk_1 = yk; yk = yk1;
        uk_1 = uk;
    }
    rms = (float)sqrt((double)rms/rcl_pts);
    printf("rms_recall = %f\n", rms);
}

savewt ()                              /* save weights */
{
    int i, j;
    FILE *fptr;
    fptr = fopen("wt.fil", "w");
    for (i=0; i<n1; i++)
        fprintf(fptr, "%f\n", bias1[i]);
    for (j=0; j<n1; j++)
    {
        for (i=0; i<n0; i++)
            fprintf(fptr, "%f ", weight1[j][i]);
        fprintf (fptr, "\n");
    }
    for (j=0; j<n2; j++)
    {
        for (i=0; i<n1; i++)
            fprintf (fptr, "%f ", weight2[j][i]);
        fprintf (fptr, "\n");
    }
    fclose (fptr);
}

readwt ()                              /* read weights */
{
    int i, j;
    FILE *fptr;
    fptr = fopen ("wt.fil", "r");
    for (i=0; i<n1; i++)
        fscanf (fptr, "%f", &bias1[i]);
    for (j=0; j<n1; j++)
```

```
        for (i=0; i<n0; i++)
            fscanf (fptr", &weight1[j][i]);
    for (j=0; j<n2; j++)
        for (i=0; i<n1; i++)
            fscanf (fptr, "%f", &weight2[j][i]);
  fclose (fptr);
}
```

Appendix E Program: Modified Elman Network for Identification

```
/***************** ElmanMod.c *********************

      modified Elman net for identification (modifiable alpha)

      ElmanMod.c: backpropagation with momentum, 1 input
      buffer layer, 1 hidden layer (linear or nonlinear), 1 output
      unit (linear). with hidden activation feedback

      program written in Microsoft Quick C (version 1.0)

      *****************************************************/

#include <conio.h>
#include <graph.h>
#include <stdlib.h>
#include <float.h>
#include <math.h>
#include <stdio.h>

#define n0 1                    /* input layer    */
#define n1 6                    /* hidden layer   */
#define n2 1                    /* output layer   */

struct videoconfig vc;

float uk_1, uk_2, yk, yk_1, yk_2;              /* define variables */
float netin0[n0+n1], netin1[n1], netin2[n2];
```

```
float bias1[n1], prdlbs1[n1];
float weight1[n1][n0+n1], weight2[n2][n1], alfa[n1];
float prdlwt1[n1][n0+n1], prdlwt2[n2][n1], prdalfa[n1];
float actv1[n1], new_ct[n1], old_ct[n1], new_gama[n1], old_gama[n1],
netout;
float lcoef, motum, rms;
unsigned seed_int, seed_u;
int epochs;
int datapoints = 400;
int recall_points = 100;
float init_wt_scale = 10.0;
float A1, A2, A3, B1, B2, D, w, a, b, gain;
float xxo, yyo;
float timegrph, ukgrph, ykgrph, netoutgrph;
char ch;

initialise ()                                    /* initialise weights */
{
  int i,j;
  srand (seed_int);
  for (i=0; i<n1; i++) {
    alfa[i] = (float)rand()/(RAND_MAX + 1.0);
    old_gama[i] = 0.0;
    actv1[i] = 0.0;
    old_ct[i] = 0.0;
  }
  for (j=0; j<n1; j++) {
    bias1[j] = (2.0*(float)rand()/RAND_MAX-1.0)/init_wt_scale;
    prdlbs1[j] = 0.0;
    for (i=0; i<(n0+n1); i++) {
      weight1[j][i] = (2.0*(float)rand()/RAND_MAX-1.0)/init_wt_scale;
      prdlwt1[j][i] = 0.0;
    }
  }
  for (j=0; j<n2; j++) {
    for (i=0; i<n1; i++) {
      weight2[j][i] = (2.0*(float)rand()/RAND_MAX-1.0)/init_wt_scale;
      prdlwt2[j][i] = 0.0;
    }
  }
}
```

```
  }

comptout ()                                    /* forward propagation */
{
  register int i, j;
  float ea, eb;
  for (j=0; j<n1; j++) {
    netin1[j] = bias1[j];
    for (i=0; i<(n0+n1); i++)
        netin1[j] += weight1[j][i] * netin0[i];
/*    ea = (float)exp((double)netin1[j]);
    eb = (float)exp((double)(-1.0)*netin1[j]);*/
    actv1[j] = netin1[j];  /*(ea - eb)/(ea + eb);*/
  }
  for (j=0; j<n2; j++) {
    netin2[j] = 0.0;
    for (i=0; i<n1; i++)
        netin2[j] += weight2[j][i] * actv1[i];
    netout = netin2[j];
  }
}

comptwt ()                                      /* update weights */
{
  register int i, j;
  float error1[n1], error2[n1], sum1[n1], sum2[n1];
  float delwt1[n1][n0+n1], delwt2[n2][n1], delalfa[n1];
  float delbs1[n1];
  for (j=0; j<n2; j++) {
    error2[j] = yk - netout;
    for (i=0; i<n1; i++)
    delwt2[j][i] = lcoef * error2[j] * actv1[i] + motum * prdlwt2[j][i];
  }
  for (j=0; j<n1; j++) {
    sum1[j] = 0.0;
    for (i=0; i<n2; i++)
        sum1[j] += error2[i] * weight2[i][j];
    error1[j] = sum1[j];   /*(1.0+actv1[j])*(1.0-actv1[j])*sum1[j];*/
    delbs1[j] = lcoef * error1[j] + motum * prdlbs1[j];
    for (i=0; i<(n0+n1); i++)
```

```
            delwt1[j][i] = lcoef*error1[j]*netin0[i] + motum*prdlwt1[j][i];
     }
   for (j=0; j<n1; j++)
      new_gama[j] = alfa[j] * old_gama[j] + old_ct[j];
   for (j=0; j<n1; j++) {
      sum2[j] = 0.0;
      for (i=0; i<n1; i++)
            sum2[j] += error1[i] * weight1[i][j];
      error2[j] = sum2[j];
   }
      for (i=0; i<n1; i++)
            delalfa[i] = lcoef*error2[i]*new_gama[i] + motum*prdalfa[i];
   for (j=0; j<n1; j++) {
      alfa[j] += delalfa[j];
      prdalfa[j] = delalfa[j];
      bias1[j] += delbs1[j];
      prdlbs1[j] = delbs1[j];
      for (i=0; i<(n0+n1); i++) {
            weight1[j][i] += delwt1[j][i];
            prdlwt1[j][i] = delwt1[j][i];
      }
   }
   for (j=0; j<n2; j++) {
      for (i=0; i<n1; i++) {
            weight2[j][i] += delwt2[j][i];
            prdlwt2[j][i] = delwt2[j][i];
      }
   }
  }
}

learn ()                                      /* train net */
{
  register int i, j, k;
  for (k=0; k<epochs; k++) {
     _clearscreen (_GCLEARSCREEN);
     _setcolor (4);
     _rectangle (_GBORDER, 0, 200, datapoints, -200);
     printf ("\nepoch %d", k);
     printf (" lcoef = %f", lcoef);
     rms = (float)sqrt((double)rms/datapoints);
```

```
printf("   rms = %f", rms);
yk_1 = yk_2 = uk_2 = 0.0;
for (i=0; i<n1; i++) {
     actv1[i] = 0.0;
     old_ct[i] = 0.0;
}
srand(seed_u);
timegrph = 0.0;  rms = 0.0;
for (j=0; j<datapoints; j++)  {
     uk_1 = gain*(2.0*(float)rand()/RAND_MAX -1.0);
     yk = A1*yk_1 + A2*yk_2 + B1*uk_1 + B2*uk_2;
     for (i=0; i<n1; i++)
       new_ct[i] = actv1[i]  + /*0.6*/ alfa[i]*old_ct[i];
     for (i=0; i<n1; i++)
       netin0[i] = new_ct[i];
     netin0[n1+n0-1] = uk_1;
     comptout ();
     ykgrph = (-1.0)*yk*100.0;
     _setcolor (2);
     _setpixel (timegrph, ykgrph);
     netoutgrph = (-1.0)*netout*100.0;
     _setcolor (7);
     _setpixel (timegrph, netoutgrph);
     comptwt ();
     rms += (yk - netout)*(yk - netout);
     yk_2 = yk_1; yk_1 = yk;
     uk_2 = uk_1;
     for (i=0; i<n1; i++) {
       old_ct[i] = new_ct[i];
       old_gama[i] = new_gama[i];
     }
     timegrph += 1.0;
  }

     if (k==(epochs-1)) {
        printf ("\nepoch %d", k+1);
        printf (" lcoef = %f", lcoef);
        rms = (float)sqrt((double)rms/datapoints);
        printf("   rms = %f", rms);
     }

}
```

```
}

recall ()                              /* test net */
{
    int i, k;
    FILE *fptr;
    _clearscreen (_GCLEARSCREEN);
    _setcolor (4);
    _rectangle (_GBORDER,0,200,recall_points*4.0,-200);
    fptr = fopen("20m3c.st2", "w");
    yk_1 = yk_2 = uk_2 = 0.0;
    timegrph = 0.0;  rms = 0.0;
    for (i=0; i<nl; i++) {
        old_ct[i] = 0.0;
        actvl[i] = 0.0;
    }

    for (k=0; k<recall_points; k++)
    {
        uk_1 = gain;
        for (i=0; i<nl; i++)
          new_ct[i] = actvl[i] + /*0.6*/ alfa[i]*old_ct[i];
          for (i=0; i<nl; i++)
            netin0[i] = new_ct[i];
        netin0[nl] = uk_1;
        comptout ();
        yk = A1*yk_1 + A2*yk_2 + B1*uk_1 + B2*uk_2;
        fprintf(fptr, "%d    %f    %f\n", k, yk, netout);
        rms += (yk - netout)*(yk - netout);
        timegrph = k*4.0;
        ykgrph = (-1.0)*yk*100.0;
        _setcolor (2);
        _setpixel (timegrph, ykgrph);
        netoutgrph = (-1.0)*netout*100.0;
        _setcolor (7);
        _setpixel (timegrph, netoutgrph);
        yk_2 = yk_1; yk_1 = yk;
        uk_2 = uk_1;
        for (i=0; i<nl; i++)
          old_ct[i] = new_ct[i];
```

```
    }
    rms = (float)sqrt((double)rms/recall_points);
    printf("rms error = %f\n", rms);
    fclose (fptr);
}


savewt ()                                    /* save weights */
{
    int i, j;
    FILE *fptr;
    fptr = fopen("wt.fil", "w");
    for (i=0; i<n1; i++)
        fprintf(fptr, "%f", bias1[i]);
    fprintf(fptr, "\n\n");
    for (j=0; j<n1; j++) {
        for (i=0; i<(n0+n1-1); i++)
            fprintf(fptr, "%f ", weight1[j][i]);
        fprintf (fptr, "\n");
    }
    fprintf(fptr, "\n");
    for (j=0; j<n1; j++)
        fprintf(fptr, "%f ", weight1[j][n0+n1-1]);
    fprintf(fptr, "\n\n");
    for (j=0; j<n2; j++) {
        for (i=0; i<n1; i++)
            fprintf (fptr, "%f ", weight2[j][i]);
        fprintf (fptr, "\n");
    }
    fprintf(fptr, "\n");
    for (i=0; i<n1; i++)
        fprintf (fptr, "%f ", alfa[i]);
    fclose (fptr);
}


readwt ()                        /* read weights */
{
    int i, j;
    FILE *fptr;
    fptr = fopen ("wt.fil", "r");
    for (i=0; i<n1; i++)
```

```
            fscanf (fptr, "%f", &bias1[i]);
     for (j=0; j<n1; j++)
          for (i=0; i<(n0+n1-1); i++)
               fscanf (fptr, "%f",  &weight1[j][i]);
     for (j=0; j<n1; j++)
          fscanf (fptr, "%f ", &weight1[j][n0+n1-1]);
     for (j=0; j<n2; j++)
          for (i=0; i<n1; i++)
               fscanf (fptr, "%f", &weight2[j][i]);
     for (i=0; i<n1; i++)
               fscanf (fptr, "%f", &alfa[i]);
     fclose (fptr);
}


main ()
{
  _setvideomode (_VRES16COLOR);
  _getvideoconfig (&vc);
  a = 1.0; w = 2.0*3.14159/2.5;
  gain = (a*a+w*w)/w;
  A1 = 1.752821; A2 = (-0.818731); B1 = 0.011698; B2 = 0.010942;
  xxo = 60;
  yyo = vc.numypixels/2.0;
  _setlogorg (xxo,yyo);
  printf("\nlearn ? (y/n)");
  ch = getche();
  if (ch == 'y') {
     printf("\n\nlearning coefficient ?");
     scanf("%f", &lcoef);
     printf("\nmomentum ?");
     scanf("%f", &motum);
     printf("\nseed_u ?");
     scanf("%d", &seed_u);
     printf("\nepochs ?");
     scanf("%d", &epochs);
     printf("\ninitialise or readwt ? (i/r)");
     ch = getche();
     if (ch == 'i') {
          printf("\nseed_int ?");
          scanf("%d", &seed_int);
```

```
        initialise ();
    }
    else if (ch == 'r')
        readwt ();
    learn ();
    savewt ();
    printf("\nlcoef=%f", lcoef);
    printf(" motum=%f", motum);
    printf(" seed_int=%d", seed_int);
    printf("\nseed_u=%d", seed_u);
    printf(" data points per epoch = %d", datapoints);
    printf(" epochs = %d\n", epochs);
    printf ("\nrecall now? (y/n)");
    ch = getche ();
    recall ();
    }
    else {
        printf ("\nrecall now? (y/n)");
        ch = getche ();
        if (ch == 'y') {
            readwt ();
            recall ();
        }
    }
    printf ("exit? (y/n)\n");
    ch = getche ();
    if (ch = 'y') {
      _clearscreen (_GCLEARSCREEN);
      _setvideomode (_DEFAULTMODE);
    }
}
```

Appendix F   Program: GMDH Network for Prediction

```
/****************** GMDH.C ***********************

              GMDH network - Adalines with nonlinear
              preprocessors - trained by Widrow-Hoff
              learning rule, used for prediction

              the training and selection data are taken
              alternatively from the available data

              program written in Microsoft Quick C (version 1.0)

****************************************************/


#include <stdlib.h>
#include <float.h>
#include <math.h>
#include <stdio.h>

#define ReDN 209        /* maximum input-output samples */
#define ML 4            /* Maximum Layers */
#define MP 60           /* Maximum PEs Per Layer */
#define ADC 6           /* channels in Adaline */
#define AD_In 2         /* inputs to each Adaline */


int NetInNo, NetInNoO, TrSeDN;              /* define variables */
float Alpha;
```

```
float Mean_y, Var_y, Dev_y;
float RDy[ReDN], RDy_test[ReDN], L_Out_t1[ReDN], Net_Out[ReDN];
int      i_No[ML][MP],      j_No[ML][MP],      i_No_Se[ML][MP],
j_No_Se[ML][MP];
int L_No, P_No[ML], Ready_L[ML], PMinL, PMinL_1;
float L_Out[ML][MP], MSEP[MP], MSEPMin, MinMSEL, MinMSEL_1;
float W[ML][MP][ADC], x[MP][ADC], PEIn[MP][AD_In];
FILE *fp, *fp1;


Read_Data()                              /* read data */
{
  int i;
  fp = fopen("exampl.dat", "r");
  for (i=0; i<ReDN; i++)
    fscanf (fp, "\n%f", &RDy[i]);
  fclose(fp);
  printf("\n%d data you just read from the disk", ReDN);
}


Show_Data()                              /* desplay data */
{
  int i;
  printf("\n\nThe data set read from disk is:");
  for (i=0; i<ReDN; i++)
    printf ("\n%f", RDy[i]);
}


MV_Proc()              /* preprocess data */
{
  int i;
  char opt[3];
  Mean_y = Var_y = 0.0;
  for (i=0; i<ReDN; i++)
    Mean_y += RDy[i];
  printf("\tMean_y = %f\n", Mean_y /= ReDN);
  for (i=0; i<ReDN; i++)
    Var_y += (RDy[i]-Mean_y)*(RDy[i]-Mean_y);
  printf("\tVar_y = %f\n", Var_y /= ReDN);
  printf("\tDev_y = %f\n", Dev_y = (float)sqrt((double)Var_y));
  printf("\n\nDo you like to proceed (type y) ? ");
```

```
        scanf("%s", opt);
        for (i=0; i<ReDN; i++)
          RDy[i] = (RDy[i] - Mean_y)/Dev_y;
        printf("\nThe processed data set is:\n");
        for (i=0; i<ReDN; i++)
          printf("\n%f", RDy[i]);
        fp = fopen("exampl.scl", "w");
        for (i=0; i<ReDN; i++)
          fprintf (fp, "\n%f", RDy[i]);
        fclose(fp);
}


Init_Wt()                               /* initialise weights */
{
  int i, j;
  for (i=0; i<P_No[L_No]; i++)
   for (j=0; j<ADC; j++)
       W[L_No][i][j] = 0.0;
}


float Adaline(int d, int c, float a, float b) /* calculate Adaline output */
{
    int i; float outvl;
    x[c][0] = 1.0;
    x[c][1] = a;
    x[c][2] = a*a;
    x[c][3] = a * b;
    x[c][4] = b * b;
    x[c][5] = b;
    for (i=0; i<ADC; i++)
        outvl += W[d][c][i]*x[c][i];
    return (outvl);
}


Train_Net()                     /* train net */
{
        char opt2[1];
        int i, j, k, p, stop_tr[MP], stop_cnt;
        unsigned long epoch;
        float MSEPOld[MP], x_Len[MP];
```

```
MinMSEL = RAND_MAX;
L_No = 0;
P_No[0] = NetInNo;
for (i=1; i<=ML; i++)
  Ready_L[i] = 0;
do
{
  L_No++;
  printf("\n\nStart training layer %d now (type y) ? ", L_No);
  scanf("%s", opt2);
  printf("\n\nL_No = %d\n", L_No);
  MinMSEL_1 = MinMSEL;
  PMinL_1 = PMinL;
  stop_cnt = 0;
  P_No[L_No] = 0;
  for (i=0; i<P_No[L_No-1]; i++)
    for (j=i+1; j<P_No[L_No-1]; j++)
    {
        i_No[L_No][P_No[L_No]] = i;
        j_No[L_No][P_No[L_No]] = j;
        P_No[L_No]++;
    }
      printf("\n\nNumber of PEs in layer %d is: %d", L_No,
P_No[L_No]);
      printf("\n\n\t**** Start training ****\n");
      for (p=0; p<P_No[L_No]; p++)
      {
        MSEPOld[p] = RAND_MAX;
        stop_tr[p] = 0;
      }
      Init_Wt();
      epoch = 0;
      do
      {
        for (k=NetInNoO; k<TrSeDN; k+=2)
        {
            for (i=0; i<NetInNo; i++)
              L_Out[0][i] = RDy[k-1-i];
            for (j=1; j<=L_No; j++)
            {
```

```
for (p=0; p<P_No[j]; p++)
{
    if (Ready_L[j] == 0)
    {
      PEIn[p][0] = L_Out[j-1][i_No[j][p]];
      PEIn[p][1] = L_Out[j-1][j_No[j][p]];
    }
    else
    {
      PEIn[p][0] = L_Out[j-1][i_No_Se[j][p]];
      PEIn[p][1] = L_Out[j-1][j_No_Se[j][p]];
    }
    L_Out[j][p]    =    Adaline(j,    p,    PEIn[p][0],
PEIn[p][1]);
}
}
for (p=0; p<P_No[L_No]; p++)
{
  if (stop_tr[p] != 1)
  {
      x_Len[p] = 0.0;
      for (i=0; i<ADC; i++)
        x_Len[p] += x[p][i]*x[p][i];
      for (i=0; i<ADC; i++)
          W[L_No][p][i]  += Alpha*x[p][i]*(RDy[k] -
L_Out[L_No][p])/x_Len[p];
  }
          MSEP[p]+=(RDy[k]-L_Out[L_No][p])*(RDy[k]-
L_Out[L_No][p]);
  }
}
for (p=0; p<P_No[L_No]; p++)
    MSEP[p] /= ((TrSeDN/2) - NetInNoO);
epoch++;
for (p=0; p<P_No[L_No]; p++)
{
    if ((MSEP[p]>=MSEPOld[p])&&(stop_tr[p]==0))
    {
       stop_tr[p] = 1;
       stop_cnt++;
```

```
                }
                MSEPOld[p] = MSEP[p];
            }
            for (p=0; p<P_No[L_No]; p++)
            {
                MSEP[p] = 0.0;
                printf("\n\nstop_tr[%d] = %d", p, stop_tr[p]);
            }
            printf("\n\nepoch = %lu, stop_cnt = %d\n\n", epoch, stop_cnt);
        } while (stop_cnt < P_No[L_No]);
        Select_f(L_No);
        printf("\nMSEPMin = %f", MSEPMin);
        MinMSEL = MSEPMin;
    } while ((MinMSEL <= MinMSEL_1)&&(P_No[L_No] != 1));
    if ((P_No[L_No] == 1)&&(MinMSEL <= MinMSEL_1))
    {
        printf("\nThe present layer has only 1 output unit");
        printf("\nLayer %d will be the output layer\n", L_No);
    }
    else
    {
        printf("\nThe minimum error in new layer is increasing\n");
        printf("\nLayer %d will be the output layer\n", L_No-1);
    }
    printf("\nEnd of training");
}


Select_f(int J)                          /* selection */
{
    float ratio[MP], threshld;
    int i, j, p, k;  char opt1, opt[1];
    printf("\n\nStart selection for layer %d now (type y) ? ", J);
    scanf("%s", opt);
    if (opt1 = ((opt[0] == 'y') || (opt[0] == 'Y')))
    {
        MSEPMin = RAND_MAX;
        printf("\n\nPE number before selection is: %d", P_No[J]);
        for (p=0; p<P_No[J]; p++)
            MSEP[p] = 0.0;
        for (k=NetInNoO; k<TrSeDN; k++)
```

```
{
    for (i=0; i<NetInNo; i++)
      L_Out[0][i] = RDy[k-1-i];
    for (j=1; j<=J; j++)
    {
      for (p=0; p<P_No[j]; p++)
      {
          if (Ready_L[j] == 0)
          {
              PEIn[p][0] = L_Out[j-1][i_No[j][p]];
              PEIn[p][1] = L_Out[j-1][j_No[j][p]];
          }
          else
          {
              PEIn[p][0] = L_Out[j-1][i_No_Se[j][p]];
              PEIn[p][1] = L_Out[j-1][j_No_Se[j][p]];
          }
          L_Out[j][p] = Adaline(j, p, PEIn[p][0], PEIn[p][1]);
      }
    }
    if (((k-NetInNoO)%2)!=0)
      for (p=0; p<P_No[J]; p++)
        MSEP[p] += (RDy[k]-L_Out[J][p])*(RDy[k]-L_Out[J][p]);
}
for (p=0; p<P_No[J]; p++)
{
    MSEP[p] /= (TrSeDN/2);
    printf("\nAv_MSEP[%d] = %f", p, MSEP[p]);
    if ((MSEPMin >= MSEP[p])&&(MSEP[p] != 0.0))
    {
      MSEPMin = MSEP[p];
      PMinL = p;
    }
}
printf("\nPMinL = %d, MSEPMin = %f", PMinL, MSEPMin);
printf("\n\nPlease give your threshld to eliminate 'bad' PEs: ");
scanf("%f", &threshld);
j = 0;
printf("\nThe i and j numbers related with the selected PEs are: ");
for (p=0; p<P_No[J]; p++)
```

```
     {
         ratio[p] = MSEP[p]/MSEPMin;
         if (ratio[p] <= threshld)
         {
             if (ratio[p] == 1.0)
                 PMinL = j;
                 i_No_Se[J][j] = i_No[J][p];   printf("\nP_No = %d:
i_No[%d][%d] = %d", p, J, p, i_No[J][p]);
                 j_No_Se[J][j] = j_No[J][p]; printf("   j_No[%d][%d] = %d",
J, p, j_No[J][p]);
             j++;
         }
     }
     P_No[J] = j;
     printf("\nAfter selection: P_No[%d] = %d, PMinL = %d\n", J,
P_No[J], PMinL);
     Ready_L[J] = 1;
     printf("\nEnd of selection\n");
   }
}


Save_Net()                     /* save net */
{
     int i, j, k;
     fp = fopen("net.1", "w");
     if ((P_No[L_No] == 1)&&(MinMSEL <= MinMSEL_1))
     {
         fprintf(fp, "\nNetInNo=%d\tL_No=%d\tPMinL=%d", NetInNo,
L_No, PMinL);
         fprintf(fp, "\nPE %d gives the smallest error in layer %d\n",
PMinL, L_No);
     }
     else
     {
         L_No--;
         fprintf(fp, "\nNetInNo = %d\tL_No = %d\tPMinL = %d",
NetInNo, L_No, PMinL_1);
         fprintf(fp, "\nPE %d gives the smallest error in layer %d\n",
PMinL_1, L_No);
     }
```

```
for (i=1; i<=L_No; i++)
{
  fprintf(fp, "\nPEs in layer %d = %d", i, P_No[i]);
   fprintf(fp, "\nThey are connected to the following PEs in layer
%d", (i-1));

    for (j=0; j<P_No[i]; j++)
    {
      fprintf(fp, "\nP_No[%d] = %d\t", j, i_No_Se[i][j]);
      fprintf(fp, "%d", j_No_Se[i][j]);
    }
    fprintf(fp, "\n");
}
fprintf(fp, "\nWeights are:\n");
for (i=1; i<=L_No; i++)
{
  fprintf(fp, "\nlayer %d\n", i);
  for (j=0; j<P_No[i]; j++)
  {
    fprintf(fp, "PE: %d\n", j);
    for (k=0; k<ADC; k++)
        fprintf(fp, "%f ", W[i][j][k]);
    fprintf(fp, "\n");
  }
}
fclose(fp);
fp = fopen("net.1a", "w");
if ((P_No[L_No] == 1)&&(MinMSEL <= MinMSEL_1))
    fprintf(fp, "\n%d\t%d\t%d", NetInNo, L_No, PMinL);
else
    fprintf(fp, "\n%d\t%d\t%d", NetInNo, L_No, PMinL_1);

for (i=1; i<=L_No; i++)
{
  fprintf(fp, "\n\n%d", P_No[i]);
  for (j=0; j<P_No[i]; j++)
  {
    fprintf(fp, "\n%d\t", i_No_Se[i][j]);
    fprintf(fp, "%d", j_No_Se[i][j]);
  }
```

```
        fprintf(fp, "\n");
      }
      for (i=1; i<=L_No; i++)
      {
        fprintf(fp, "\n");
        for (j=0; j<P_No[i]; j++)
        {
          for (k=0; k<ADC; k++)
              fprintf(fp, "%f ", W[i][j][k]);
          fprintf(fp, "\n");
        }
      }
      fclose(fp);
}


Read_Net()                              /* read saved net */
{
      int i, j, k;
      fp = fopen("net.1a", "r");
      fscanf(fp, "%d %d %d", &NetInNo, &L_No, &PMinL);
      for (i=1; i<=L_No; i++)
      {
        fscanf(fp, "%d", &P_No[i]);
        for (j=0; j<P_No[i]; j++)
        {
          fscanf(fp, "%d", &i_No_Se[i][j]);
          fscanf(fp, "%d", &j_No_Se[i][j]);
        }
      }
      for (i=1; i<=L_No; i++)
      {
        for (j=0; j<P_No[i]; j++)
        {
          for (k=0; k<ADC; k++)
          {
              fscanf(fp, "%f", &W[i][j][k]);
          }
          printf("\n");
        }
      }
```

```
        fclose(fp);
}


Test_Net()                               /* test net */
{
    int i, j, p, k;
    fp = fopen("test.1", "w");
    j = 0;
    for (p=0; p<P_No[L_No]; p++)
        MSEP[p] = 0.0;
    for (k=NetInNoO; k<ReDN; k++)
    {
        for (i=0; i<NetInNo; i++)
          L_Out[0][i] = RDy[k-1-i];
        for (j=1; j<=L_No; j++)
        {
          for (p=0; p<P_No[j]; p++)
          {
             PEIn[p][0] = L_Out[j-1][i_No_Se[j][p]];
             PEIn[p][1] = L_Out[j-1][j_No_Se[j][p]];
             L_Out[j][p] = Adaline(j, p, PEIn[p][0], PEIn[p][1]);
          }
        }
        RDy_test[k] = Dev_y*RDy[k] + Mean_y;
        L_Out_t1[k] = Dev_y*L_Out[L_No][PMinL] + Mean_y;
        fprintf(fp, "\n%d\t%f\t%f", k, RDy_test[k], L_Out_t1[k]);
        MSEP[PMinL] += (RDy_test[k] - L_Out_t1[k])*(RDy_test[k] -
L_Out_t1[k]);
    }
    fclose(fp);
    MSEP[PMinL] /= (ReDN - NetInNoO);
    printf("Minimum(1):      MSEP[%d]      =      %f",      PMinL,
(float)sqrt((double)MSEP[PMinL]));

    for (k=0; k<ReDN; k++)
        Net_Out[k] = RDy[k];
    fp = fopen("fntest.1a", "w");
    j = 0;
    for (p=0; p<P_No[L_No]; p++)
        MSEP[p] = 0.0;
```

```
    for (k=NetInNoO; k<ReDN; k++)
    {
        for (i=0; i<NetInNo; i++)
          L_Out[0][i] = Net_Out[k-1-i];
        for (j=1; j<=L_No; j++)
        {
          for (p=0; p<P_No[j]; p++)
          {
             PEIn[p][0] = L_Out[j-1][i_No_Se[j][p]];
             PEIn[p][1] = L_Out[j-1][j_No_Se[j][p]];
             L_Out[j][p] = Adaline(j, p, PEIn[p][0], PEIn[p][1]);
          }
        }
        if (k >= TrSeDN)
          Net_Out[k] = L_Out[L_No][PMinL];
        RDy_test[k] = Dev_y*RDy[k] + Mean_y;
        L_Out_t1[k] = Dev_y*L_Out[L_No][PMinL] + Mean_y;
        fprintf(fp, "\n%d\t%f\t%f", k, RDy_test[k], L_Out_t1[k]);
        MSEP[PMinL] += (RDy_test[k] - L_Out_t1[k])*(RDy_test[k] -
L_Out_t1[k]);
    }
    fclose(fp);
    MSEP[PMinL] /= (ReDN - NetInNoO);
    printf("\nMinimum(m):     MSEP[%d]     =     %f",     PMinL,
(float)sqrt((double)MSEP[PMinL]));
}

main ()
{
  char option, opt[1];
  Read_Data();
  Show_Data();
  printf("\n\nDo you like to use original data (o) or processed data (p)? ");
  scanf("%s", opt);
  if (option = ((opt[0] == 'p') || (opt[0] == 'P')))
    MV_Proc();
  printf("\n\nHow many data for the training and selection sets ? ");
  scanf("%d", &TrSeDN);
  printf("\n\nHow many inputs do you have for the network ? ");
  scanf("%d", &NetInNo);
```

```
printf("\nNetInNo = %d", NetInNo);
NetInNoO = NetInNo;
printf("\n");
printf("\nAre you tRaining or tEsting (R or E) ? ");
scanf("%s", opt);
if (option = (opt[0]=='r') || (opt[0]=='R'))
{
        printf("\nPlease give a learning rate (between 0.1 and 1.0) ");
        scanf("%f", &Alpha);
        printf("\nAlpha for this training session is = %f", Alpha);
        Train_Net();
        Save_Net();
}
else if (option = (opt[0]=='e') || (opt[0]=='E'))
{
        Read_Net();
        Test_Net();
}
do
{
        printf("\n\nExit ? (type y) ");
        scanf("%s", opt);
} while ((opt[0]!='y') || (opt[0]=='Y'));
}
```

Author Index

Subject Index